Transistors!

New Era Electronics: A Lecture Notes Series

Series Editors: Vijay Raghunathan *(Purdue University, USA)*
Muhammad Ashraf Alam *(Purdue University, USA)*
Mark S Lundstrom *(Purdue University, USA)*

Published

Vol. 1   *Transistors!*
by Mark S Lundstrom

New Era Electronics: a Lecture Notes Series ▪ Volume 1

Transistors!

Mark Lundstrom

Purdue University, USA

World Scientific

NEW JERSEY · LONDON · SINGAPORE · BEIJING · SHANGHAI · HONG KONG · TAIPEI · CHENNAI · TOKYO

Published by

World Scientific Publishing Co. Pte. Ltd.

5 Toh Tuck Link, Singapore 596224

USA office: 27 Warren Street, Suite 401-402, Hackensack, NJ 07601

UK office: 57 Shelton Street, Covent Garden, London WC2H 9HE

Library of Congress Control Number: 2022950370

British Library Cataloguing-in-Publication Data
A catalogue record for this book is available from the British Library.

New Era Electronics: A Lecture Notes Series — Vol. 1
TRANSISTORS!

ISBN 978-981-126-726-0 (hardcover)
ISBN 978-981-126-768-0 (paperback)
ISBN 978-981-126-727-7 (ebook for institutions)
ISBN 978-981-126-728-4 (ebook for individuals)

For any available supplementary material, please visit
https://www.worldscientific.com/worldscibooks/10.1142/13168#t=suppl

Desk Editor: Joseph Ang

Typeset by Stallion Press
Email: enquiries@stallionpress.com

To the transistor pioneers

and those who followed in their footsteps.

Preface

"During the last two decades electronics has become one of the most exciting industries of all times... This new electronics lives close to the frontiers of science, and requires a high level of technical competence. It grows by the development of new products. It is characterized by the **transistor** and other solid state electronic devices..."
Frederick Terman, 16th National Electronics Conf., Oct. 16, 1960.

The publication of these lecture notes coincides with the 75th anniversary of the invention of the transistor in 1947. When Terman gave his speech, transistors were still serving niche markets, but he saw where electronics would go. Since 1960, the world has been transformed by the continuous increase in the capabilities of electronic systems and the continuous decrease in cost enabled by transistors (and integrated circuits (ICs), which were invented about the time Terman gave his speech). It has been argued that the transistor was the most important invention of the 20th century, and it now seems certain that the impact of transistors in integrated circuit chips will be even greater in the 21st century. Transistors are the basic components of electronic systems — the "atoms" of electronics. My goal in these lectures is to convey in a physically sound but simple way the basic operating principles of these remarkable devices.

Over the past 75 years, many excellent books on transistors have been written — why write another one? The reason is that those books reflect the times in which they were written, and today we are entering a new era of electronics. Making transistors smaller has been the driving force for progress in electronics since the invention of the IC. As transistors were down-sized, new physical effects came into play and had to be understood, modeled, and incorporated into textbooks. The result is that books often include material that is still highly relevant and essential along with discussions of topics and approaches that are no longer of central importance. Extensive discussions of what once were critical factors for transistors can cause confusion about what is important today.

Today's leading-edge CMOS transistors are about as small as they will get. We now have a simple, clear, very physical understanding of how these devices function, but it has not yet gotten into our textbooks. As we begin a new era in which making transistors smaller will no longer be a major driving force for progress, it is time to look back at what we have learned in transistor research and convey as simply and clearly as possible the essential physics of the device that makes modern electronics possible. That is the goal of these lectures.

Readers familiar with transistors will see much that is familiar here, but also several things that are not typically found in textbooks. For example, rather than beginning with bulk, planar MOSFETs and treating fully-depleted MOSFETs as an "advanced topic," as is typical, I begin with fully-depleted MOSFETs, which are both more relevant to modern logic transistor technology and pedagogically easier to discuss. Another example is the treatment of today's logic MOSFETs with 10-nm scale channel lengths. Beginning with the scattering dominated models for longer channel lengths makes the physics of these devices look unnecessarily complicated. Beginning with a simple ballistic model for a MOSFET and then adding in some scattering is a clearer approach.

The semiconductor physicist, Herbert Kroemer, said it well: "If, in discussing a semiconductor problem, you cannot draw an Energy Band Diagram, this shows that you don't know what you are talking about. If you can draw one, but don't, then your audience won't know what you are talking about" (2000 Nobel Lecture). There is little discussion of energy band diagrams in most transistor textbooks, but the lecture on the energy band treatment of the MOSFET is the most important one in this volume because if you understand MOSFETs in terms of energy band diagrams, everything else is just details.

The last lecture is on transistor reliability, a topic rarely discussed in a general, introductory textbook. At the beginning of semiconductor technology, some thorny reliability problems had to be solved just to make the technology viable. Today, IC manufacturers must assure their customers that the chips will function reliably for 10 years or so. As transistor technology evolved over many years, the dominant transistor wear-out mechanisms continually changed, and this is likely to continue as the applications for

ICs and the environments in which they operate continue to diversity. A basic understanding of the intrinsic wear-out processes for transistors is becoming essential for semiconductor technologists.

These brief notes focus on the most common type of transistor, the Metal-Oxide-Semiconductor Field-Effect Transistor (MOSFET), but the fundamental operating principle of almost all types of transistors is the same as that of a MOSFET. A few, important principles are discussed; details are omitted. For those who simply want to understand how transistors work, this treatment should suffice. For those pushing the frontiers of transistor science and technology, these lectures provide an overall framework that can be filled in with problem-specific deep dives. Instructors using these lecture notes will undoubtedly notice some omissions, but rather than including every topic that any instructor might feel necessary, I have chosen to keep the volume short and provide instructors with an opportunity to supplement the notes where they feel it is needed. Depending on how they are supplemented, the lectures could be used at the undergraduate or graduate levels.

To follow these lectures, readers need only a basic understanding of semiconductor physics. Familiarity with transistors and electronic circuits is helpful, but not assumed. The treatment here is a distillation of the essentials from a more comprehensive treatment by the author [1], which discusses in more detail some topics that are only touched upon here. For more comprehensive and more conventional treatments of transistors, see [2, 3]. I hope readers find these lecture notes a useful, succinct introduction to a device that is both scientifically interesting and technologically important.

Mark Lundstrom
Purdue University
July 31, 2022

[1] Mark Lundstrom, *Fundamentals of Nanotransistors*, Vol. 6 in *Lessons from Nanoscience: A Lecture Notes Series*, World Scientific Publishing Company, Singapore, 2018.

[2] Yannis Tsvidis and Colin McAndrew, *Operation and Modeling of the MOS Transistor*, 3rd Ed., Oxford Univ. Press, Oxford, UK, 2010.

[3] Yuan Taur and Tak Ning, *Fundamentals of Modern VLSI Devices*, 3rd Ed., Cambridge Univ. Press, Cambridge, UK, 2022.

Acknowledgments

It is a pleasure to acknowledge several people who contributed to this volume. First, thanks to World Scientific Publishing Company and our series editor, Zvi Ruder, for their support with this new lecture notes series. These lectures summarize what I've learned over many years in working with a remarkable group of students on challenging transistor research problems. I have also learned much from my colleagues beginning with Professor Supriyo Datta of Purdue University whose approach to carrier transport at the nanoscale provided a clear, simple, and sound way to understand transport in nanoscale MOSFETs. Professor Dimitri Antoniadis of the Massachusetts Institute of Technology developed a "virtual source" model of that captures the essential ideas discussed in these lectures and provided a clear, simple model useful for technology assessment and benchmarking. His careful analysis of nanoscale transistor measurements has done much to clarify my own understanding of these remarkable devices.

It is also a pleasure to acknowledge the help of Professors Jing Guo of the University of Florida, Shaloo Rakheja of the University of Illinois at Urbana-Champaign, and Sumeet Gupta of Purdue University who carefully read and critiqued the first draft. The manuscript was much improved by their efforts.

I owe special thanks to three other colleagues. Drs. James Cooper, Jr. and Dallas Morisette helped me understand the fundamentals of power MOSFETs and played a strong role in shaping and refining the lecture on power MOSFETs. Similarly, Professor M.A. Alam helped me learn the fundamentals of transistor reliability and was deeply involved in shaping and refining the last lecture in the volume. Without their extensive help, I would not have been able to include the two lectures on these important topics.

These lectures are an attempt to summarize as simply as possible the essential physics of transistors. I hope they provide a foundation for creative innovations in 21st century electronics, and I am grateful to the opportunities I've had to work on transistors over many years with so many exceptions students and colleagues.

Contents

List of Figures

A First Look at Transistors

1.1   Introduction

This first lecture describes but does not explain the *IV* characteristics of transistors. The next lecture discusses how transistors are used in digital and analog circuits and defines some *device metrics* used to assess the performance of transistors in circuits. Subsequent lectures will relate the *IV* characteristics to the underlying physics.

These lectures are about what goes on inside a transistor, but for this first lecture, we treat a transistor as an engineer's "black box" shown in Fig. 1.1. A large current flowing through terminals 1 and 2 is controlled by the voltage on (or, for some transistors, the current injected into) terminal 3. Sometimes there is a fourth terminal too. There are many kinds of transistors [1], but all transistors have three (or four) external leads like the generic one sketched in Fig. 1.1. The names given to the terminals depend on the type of transistor. The most common type of transistor is the field-effect transistor (FET). In these lectures, our focus is on a specific type of FET, the silicon Metal-Oxide-Semiconductor Field-Effect Transistor (MOSFET). A different type of FET, the High Electron Mobility Transistor

(HEMT), finds use in radio frequency (RF) applications. Bipolar junction transistors (BJTs) and heterojunction bipolar transistor (HBTs) are also used for RF applications. Although our focus is on the Si MOSFET, the same principles apply to these other transistors as well; they will be briefly discussed in Lecture 11.

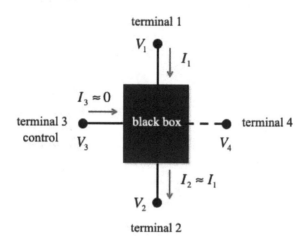

Fig. 1.1   Illustration of a transistor as a black box. The currents that flow in the four leads of the device are controlled by the voltages applied to the four terminals. Terminal 4 is special; it is present in some transistors but not in others.

The relation of the currents to the voltages is determined by the internal physics of the transistor. These lectures will relate the current vs. voltage characteristics to the underlying device physics of transistors and develop simple, analytical expressions for the *IV* characteristics.

1.2   Inside the black box

Figure 1.2 shows a scanning electron micrograph (SEM) cross section of a Si MOSFET circa 2000. The drain and source terminals (terminals 1 and 2 in Fig. 1.1) are clearly visible, as are the gate electrode (terminal 3 in Fig. 1.1) and the Si body contact (terminal 4 in Fig. 1.1). Note that the gate electrode is separated from the Si substrate by a thin, insulating layer that is less than 2 nm thick. The region between the source and the drain is called the *channel.*

Also shown in Fig. 1.2 is the schematic symbol used to represent MOS-FETs in circuit diagrams. The dashed line represents the channel between the source and drain. It is dashed to indicate that this is an *enhancement mode* MOSFET, one that is only "on" with a channel present when the magnitude of the gate voltage exceeds a critical value known as the *threshold voltage, V_T*.

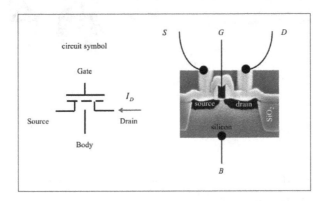

Fig. 1.2   The n-channel silicon MOSFET. Left: The circuit schematic of an enhancement mode n-channel MOSFET showing the source, drain, gate, and body contacts. The dashed line represents the channel, which is present when a large enough gate voltage is applied. Right: An SEM cross-section of a silicon MOSFET circa 2000. The source, drain, gate, silicon body, and gate insulator are all visible.

Figure 1.3 compares the cross-sectional and top-view of an n-channel, silicon MOSFET. On the left is a "cartoon" illustration of the cross-section, similar to the SEM in Fig. 1.2. In an n-channel MOSFET, the source and drain are heavily doped n-type regions, and the transistor operates by controlling the flow of electrons across the channel that separates the source and drain. On the right side of Fig. 1.3 is a top view of the same transistor. The large rectangle is the transistor itself. The squares with "X"s on the two ends of this rectangle are contacts to the source and drain regions, and the black rectangle in the middle is the gate electrode. Below the gate is the gate oxide, and under it, the silicon channel.

The channel length, L, is a critical parameter that determines the ultimate speed of the transistor (the shorter L is, the faster the ultimate speed of the transistor). The width, W, determines the magnitude of the current that flows. Circuit designers specify the lengths and widths of transistors to achieve the desired circuit performance. For a given technology, transis-

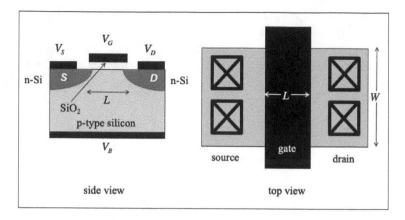

Fig. 1.3 Comparison of the cross-sectional, side view (left) and top view (right) of an n-channel, silicon MOSFET.

tors are designed to be well-behaved for channel lengths greater than some minimum channel length, which, for present day technologies, is about 10 nanometers (10 nm). The minimum channel length, the minimum contact size needed to provide low resistance contacts to the source and drain, and the minimum spacing between the contacts and the gate electrode all determine the minimum size (or "footprint") of the transistor. For the several decades, engineers have steadily shrunk the minimum channel length and other dimensions to reduce the size of transistors so that more transistors can be placed on an integrated circuit "chip." The steadily increasing number of transistors per chip is known as *Moore's Law* [2, 3].

1.3 Common source, common drain, and common gate

In circuits, transistors are usually configured to accept an input voltage and produce an output voltage. The input voltage is measured across two input terminals and the output voltage across two output terminals. For MOSFETs, the gate current is very small because of the insulator below the gate electrode. The output current is the current that flows into one of the two output terminals and out of the other. Since we only have three terminals (terminal 4 is special), one of the terminals must be connected in common to provide two input and two output terminals. Possibilities are *common source*, *common drain*, and *common gate* configurations.

Figure 1.4 shows an n-channel MOSFET connected in the common source configuration. (For N-MOSFETs, the charge carriers are negatively charged electrons. As discussed in Sec. 1.5, there are also P-MOSFETs, which rely on positive charge carriers.) In the common source case, the output current is I_D, and the output voltage is the drain to source voltage, V_{DS}. The input voltage is the gate to source voltage, V_{GS}. For MOSFETs, the DC gate current is typically very small and can often be neglected. Note that the body contact is special — when it is present, it just tunes the operating characteristics of the MOSFET.

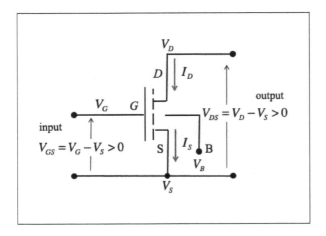

Fig. 1.4   An n-channel MOSFET configured in the common source mode. The input voltage is V_{GS}, and the output voltage, V_{DS} . The output current is I_D, and the gate current is typically negligibly small, so the DC input current is assumed to be zero.

Note the conventions for labeling the DC currents and voltages as shown in Fig. 1.4. For example, V_D is the voltage on the drain. Voltages are always measured with respect to a reference called ground. If $V_D = 1$ V, it means that the drain voltage is 1 V above the reference voltage, which is usually taken to be 0 V. Two important voltages are the voltage between the drain and the source, $V_{DS} = V_D - V_S$ and the voltage between the gate and the source, $V_{GS} = V_G - V_S$. The DC current that flows into the drain is I_D, which is nearly the same as the current that flows out of the source, I_S, because the gate and body currents are approximately zero. The directions of the currents in Fig. 4 are the directions for an n-channel MOSFET.

Our goal in this lecture is to understand the general features of transistor IV characteristics and to introduce some of the terminology used. Two types of *IV* characteristics are of interest; the first are the *output character-istics.* For the common source configuration, this is a plot of the output current, I_D, vs. the output voltage, V_{DS}, for a constant input voltage, V_{GS}. The second *IV* characteristic of interest is the *transfer characteristic.* For the common source configuration, this is a plot of the output current, I_D, as a function of the input voltage, V_{GS}, for a fixed output voltage, V_{DS}. In the remainder of this lecture, we describe the common source *IV* characteristics and define some terminologies.

1.4   IV characteristics

Figure 1.5 shows the *IV* characteristics of three devices – a resistor, an ideal current source, and a transistor.

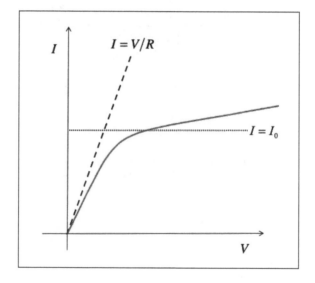

Fig. 1.5   The *IV* characteristics of a resistor (dashed line), and an ideal current course (dotted line). The solid line shows I_D vs. V_{DS} for a transistor at a fixed V_{GS}.

The dashed line in Fig. 1.5 shows a resistor for which the current is pro-portional to the voltage according to $I = V/R$, where R is the resistance in Ohms. The dotted line shows an ideal current source — the current is independent of voltage across the terminals of the current source. The solid line is a MOSFET output characteristic — a plot of I_D vs. V_{DS} for a fixed input voltage, V_{GS}. We see that for small V_{DS}, the transistor behaves like a resistor, $I_D \propto V_{DS}$. For large V_{DS}, the transistor behaves like a current source, but not an ideal current source because there is some dependence of I_D on V_{DS}. For large V_{DS}, the transistor behaves like an ideal current source in parallel with a resistor.

The output characteristics of an n-channel MOSFET are shown in Fig. 1.6. Each line in the family of characteristics corresponds to a different input voltage, V_{GS}. For small V_{DS} (the *linear* or *ohmic* region of operation), a MOSFET operates like a resistor with the resistance being given by the inverse of the slope of I_D vs. V_{DS} for small V_{DS}. As shown in Fig. 1.6, the resistance decreases as V_{GS} increases.

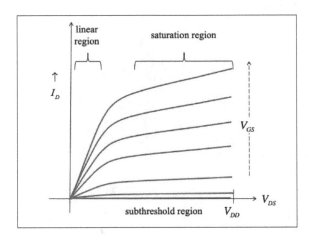

Fig. 1.6   The common source output IV characteristics of an n-channel MOSFET. The vertical axis is the current that flows between the drain and source, I_D, and the horizontal axis is the voltage between the drain and source, V_{DS}. Each line corresponds to a different gate voltage, V_{GS}. The two regions of operation, linear (or ohmic) and saturation, are also labeled.

For V_{DS} greater than a voltage known as the *drain saturation voltage*, V_{DSAT}, the MOSFET operates as a gate voltage dependent current source. This is known as the *saturation* region of operation. The magnitude of the saturation region current increases with V_{GS}.

In the saturation region, we also see that the current increases a little with increasing V_{DS}, which shows that the current source has a finite output resistance, r_o, given by the inverse of the slope of I_D vs. V_{DS} in the saturation region. Finally, for $V_{GS} < V_T$, the drain current is very small and not visible when plotted on a linear scale as in Fig. 1.6. This is the *subthreshold region.*

Figure 1.7 compares the common source output and transfer characteristics of an n-channel MOSFET. Consider fixing V_{DS} to a small value and sweeping V_{GS}. This gives the line labeled V_{DS1} in the transfer characteristics on the right. If we fix V_{DS} to a large value and sweep V_{GS}, then we get the line labeled V_{DS2} in the transfer characteristic. The transfer characteristic also shows that for $V_{GS} < V_T$, the current is very small. This is the *subthreshold region* of operation. A plot of $\log_{10}(I_D)$ vs. V_{GS} is used to resolve the current in the subthreshold region.

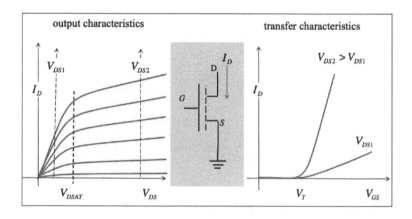

Fig. 1.7 Common source output (left) and transfer (right) characteristics of an n-channel MOSFET. The voltage, $V_{DS1} \ll V_{DSAT}$ is in the linear region and $V_{DS2} \gg V_{DSAT}$ is in the saturation region.

1.5   N and P-channel MOSFETs

In the n-channel MOSFET shown on the left in Fig. 1.8, conduction is by electrons in the conduction band. As shown on the right in Fig. 1.8, it is also possible to make the complementary device in which conduction is by holes in the valence band. In a p-channel MOSFET, the source and drain regions are heavily doped p-type, and the transistor operates by controlling hole conduction across the channel that separates the source and drain. Figure 1.8 shows sketches of traditional MOSFETs, so-called *bulk* or *planar* MOSFETs, which are built on the surface of a thick silicon wafer. For n-channel MOSFETs, the wafer is doped p-type, and for p-channel MOS-FETs, the wafer is doped n-type. Modern MOSFETs for digital electronics have a somewhat different physical structure. One of these is the so-called FinFET, which we will discuss later, but the operating principles are the same.

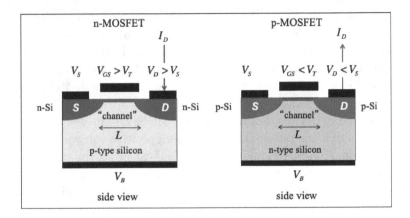

Fig. 1.8   Comparison of an n-channel MOSFET (left) and a p-channel MOSFET (right). Note that $V_{DS}, V_{GS} > 0$ for the n-channel device and $V_{DS}, V_{GS} < 0$ for the p-channel device. The drain current flows in the drain of an n-channel MOSFET and out the drain of a p-channel MOSFET. The body is often shorted to the source so that $V_B = V_S$.

It is important to note that $V_{DS} > 0$ for the n-channel device and $V_{DS} < 0$ for the p-channel device. To remember this, think of the positive voltage on the drain of the n-MOSFET as attracting negatively charged electrons from the source, so that they can flow across the channel to the drain. Similarly, think of the negative voltage on the drain of the p-MOSFET as attracting positively charged holes from the source, so that they can flow across the channel to the drain.

For the enhancement mode devices we are considering, it is also important to note that $V_{GS} > 0$ for the n-channel device and $V_{GS} < 0$ for the p-channel device. A channel for current to flow from source to drain does not exist until the appropriate gate voltage is applied. Think of the positive voltage on the gate of the n-MOSFET as attracting negatively charged electrons to the surface to produce an n-type channel that connects the source and drain. Similarly, think of the negative voltage on the gate of the p-MOSFET as attracting positively charged holes to the surface to produce a p-type channel that connects the source and drain. Finally, note that the drain current flows into the drain of the n-MOSFET (i.e. electrons flow out of the drain) and that the drain current flows out of the drain of the p-MOSFET (i.e. holes flow out of the drain). Modern electronics is largely built with CMOS (or complementary MOS) technology for which every n-channel device is paired with a p-channel device.

There is no generally agreed upon standard for drawing transistor symbols in circuit diagrams. Figure 1.9 shows how we will represent n-channel and p-channel MOSFETs. On the left is an n-channel, enhancement mode MOSFET. The dashed line between the source and drain indicates that a conduction channel does not exist until $V_{GS} = V_G - V_S > V_T > 0$. The right side of Fig. 1.9 illustrates p-channel, enhancement mode MOSFETs for which a channel does not exist unless $V_{GS} < V_T < 0$. MOSFETs are typically symmetrical devices. For an n-channel MOSFET, $V_{DS} = V_D - V_S > 0$, which means that the terminal with the more positive voltage is the drain. For a p-channel MOSFET, $V_{DS} < 0$, which means that the terminal with the more negative voltage is the drain. As shown on the far right, it is convenient to work in terms of positive voltages for p-channel transistors. Accordingly, p-channel MOSFETs are often drawn upside down, so that the source is connected to the most positive voltage, which is typically at the top. For a p-channel MOSFET to be on, $V_{SG} > |V_T| > 0$ and $V_{SD} > 0$.

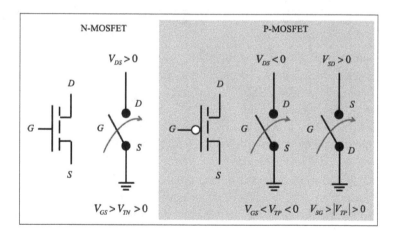

Fig. 1.9 The schematic symbols used to differentiate n- and p-channel MOSFETs. An open circle on the gate denotes a PMOS transistor. The sign of V_{GS} and V_{DS} needed to turn the transistor on is also indicated. As shown on the right, PMOS transistors are often drawn upside down with the source on the top.

1.6 Summary

In this lecture we described the *IV* characteristics of MOSFETs and defined several terms that we will often encounter:

- Threshold voltage, V_T
- Enhancement mode
- Drain saturation voltage, V_{DSAT}
- Linear, saturation, and subthreshold regions of operation
- Source, drain, channel, body
- Gate oxide
- Common source, common drain, common gate
- Output and transfer characteristics
- n- and p-channel MOSFETs
- CMOS technology

Our discussion has focussed on n-channel MOSFETs. To test your understanding, determine the signs of V_{GS} and V_{DS} in Fig. 1.4 and sketch Figs. 1.6 and 1.7 for p-channel MOSFETs.

We also briefly discussed the physical structure of a MOSFET but did not discuss what goes on inside the black box to produce the *IV* characteristics. Subsequent lectures will focus the physics that produces these *IV* characteristics, but first, in Lecture 2, we will briefly discuss how transistors are used in digital and analog circuits.

Lecture 1 Exercise: n-channel and p-channel MOSFETs

To gain familiarity with p- and n-channel MOSFETs, consider the two MOSFETs in parallel as sketched below. Assume that the threshold voltages are $V_{TN} = -V_{TP} = V_T > 0$, and that $V_{DD} > V_T$. We have not yet discussed the mathematical modeling of I_D, but you can assume that when the transistors are on, $I_D \propto (V_{GS} - V_{TN})$ for the n-channel MOSFET and and $I_D \propto |(V_{GS} - V_{TP})|$ for the p-channel MOSFET.

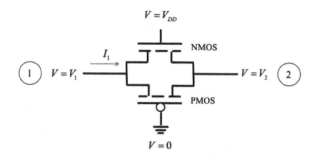

An n-channel and a p-channel MOSFET connected in parallel.

1) If $V_1 = V_{DD}$ and $V_2 = 0$ V, what is I_1?

For an NMOS transistor, the terminal with the higher voltage is the drain, so Terminal 1 is the drain and Terminal 2 is the source. The NMOS transistor is on with $V_{GS}(N) = V_{DD}$. For a p-channel MOSFET (PMOS), the terminal with the higher voltage is the source, so Terminal 1 is the source and Terminal 2 is the drain. The PMOS transistor is on with $V_{GS}(P) = -V_{DD}$. Since $V_{GS}(N) = -V_{GS}(P)$, and $V_{TN} = -V_{TP}$, the drain currents are identical. For the NMOS transistor, I_D flows into the drain, and for the PMOS it flows into the source. We conclude that $I_1 = 2I_D$.

2) If $V_1 = 0$ V and $V_2 = V_{DD}$, what is I_1?

For the NMOS transistor, Terminal 2 is now the drain and Terminal 1 is the source. The NMOS transistor is on with $V_{GS}(N) = V_{DD}$. For the PMOS transistor, Terminal 2 is now the source and Terminal 1 is the drain. The PMOS transistor is on with $V_{GS}(P) = -V_{DD}$. The drain currents are identical. For the NMOS transistor, I_D flows into the drain, and for the PMOS it flows into the source. In this case, we conclude that $I_1 = -2I_D$.

1.7   References

There are many types of transistors, for an incomplete list, see:

[1] Kwok K. Ng "A survey of semiconductor devices," *IEEE Trans, Electron Devices*, **43**, pp. 1760–1766, 1996.

For an introduction to Moore's Law, see:

[2] "Moore's law," `http://en.wikipedia.org/wiki/Moore's_law`, July 19, 2013.

[3] M. Lundstrom, "Moore's Law Forever?" an Applied Physics Perspective, *Science*, **299**, pp. 210–211, January 10, 2003.

Lecture 2

Circuits and Device Metrics

2.1 Introduction

Transistors are the basic components of electronic systems; they have two main uses. The first is for digital electronics or in applications such as switching power supplies where the transistor is a switch that is either off or on. The second is to amplify audio or radio frequency signals for applications such as audio players or in wireless communication. A few simple *device metrics* help designers understand how a transistor will perform in circuits. Our goal in this lecture is to discuss some very basic circuit considerations in order to appreciate the importance of a few key device metrics.

2.2 Digital circuits

An understanding of the CMOS inverter shown in Fig. 2.1 provides insight into digital electronics. NAND and NOR gates can be implemented with similar circuits. As shown in Fig. 2.1, a CMOS inverter consists of two MOSFETs connected in series with the source of the PMOS transistor connected to the highest voltage, V_{DD}, and the source of the NMOS transistor connected to the lowest voltage. The input voltage is applied to the two gates, and the output voltage is taken from the two drains.

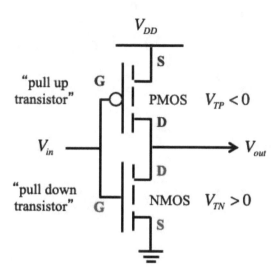

Fig. 2.1   The basic building block of Complementary MOS digital circuits — the CMOS inverter. The PMOS transistor is referred to as the *pull up transistor* because when on, it pulls the output voltage up to V_{DD}. The NMOS transistor is referred to as the *pull down transistor* because when on, it pulls the output voltage down to 0 V.

The operation of this circuit is easy to understand if we think of the transistors as switches that are either off or on. If $V_{in} = V_{DD}$ (a logic 1), then the NMOS transistor is on ($V_{GS} = V_{DD} > V_{TN}$), and the PMOS transistor is off ($V_{SG} = 0$). The output is connected through the NMOS transistor to ground, so $V_{out} = 0$, a logic 0. If $V_{in} = 0$ (a logic 0), then the NMOS transistor is off, but $V_{SG} = V_{DD} > |V_{TP}|$, so the PMOS transistor is on, and the output is connected through the PMOS transistor to V_{DD}, a logic 1. The circuit converts a logical 0 to a 1 and a logical 1 to a 0.

It is important to note that when the input is either a 0 or a 1, only a small leakage current flows because one of the two series-connected transistors is off. Significant current only flows while switching between 0 and 1. Without this property, the power dissipation of a large digital system would be excessive. Because of the large numbers of transistors in modern systems, the small leakage currents that do flow when transistors are off can add up and contribute to a significant *standby power*.

Figure 2.2 is a plot of the transfer characteristic — the output voltage vs. the input voltage. For a good CMOS inverter, the transition from a high output voltage to a low output voltage should occur over a very small range of input voltages. The slope of V_{out} vs. V_{in} at the transition point should be large. This is the voltage gain of the circuit, A_V. Note that if $|A_V| > 1$, then there are *noise margins*. The input voltage only needs to be low enough to produce an exactly correct logical 1 output; it does not have to be exactly 0 V. Similarly, the input voltage only needs to be high enough to produce an exactly correct logical 0 output; it does not need to be exactly V_{DD}. With billions and billions (and now even trillions) of transistors on a digital chip, errors would quickly accumulate without these noise margins that reset signals to their proper values. For this reason, voltage gain is necessary for digital logic.

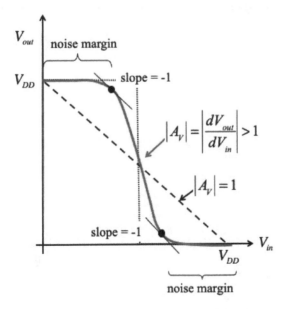

Fig. 2.2 The transfer characteristic of the CMOS inverter. The dotted line is an ideal inverter, and the solid line is the transfer characteristic of a real CMOS inverter. Voltage gain is defined at the mid-point of the switching transition. The dashed line shows that when the voltage gain has a magnitude of one, the noise margin is zero.

The next question is: "How fast can we change a 0 to a 1 or vice versa?" The switching speed will determine how fast our computer can operate. The basic ideas are illustrated in Fig. 2.3. The capacitor, C_{sw} is the *switching capacitance*, which represents the total capacitance of the gates being driven.

Consider first a 1 to 0 transition of V_{in} as shown on the left in Fig. 2.3. Prior to switching, the input voltage was high, so the NMOS transistor was on, making the output voltage zero. When the input switches low, the NMOS transistor turns off; the PMOS transistor turns on so current flows from the power supply into C_{sw} charging it up to V_{DD}. A charge of $Q = C_{sw}V_{DD}$ has been placed on C_{sw}.

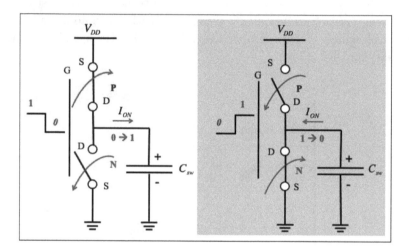

Fig. 2.3   Illustration of the switching behavior of the CMOS inverter. Left: 1 to 0 input transition, which charges C_{sw}. Right: 0 to 1 input transition, which discharges C_{sw}.

Consider next a 0 to 1 transition of V_{in} as shown on the right in Fig. 2.3. Prior to switching, the input voltage was low, so the NMOS transistor was off, but the PMOS transistor was on, so C_{sw} would have been charged up through the PMOS transistor making the output voltage V_{DD}. When the input switches high, the NMOS transistor turns on; the PMOS transistor turns off, so current flows from C_{sw} to ground through the NMOS transistor. A charge of $Q = C_{sw}V_{DD}$ has been removed from C_{sw}.

How long does it take to charge or discharge C_{sw}? A typically designed CMOS inverter is symmetrical, so the times to charge or discharge C_{sw} are the same. Since $i = dQ/dt$, we can write

$$\frac{Q}{2} = \frac{C_{sw}V_{DD}}{2} = \int_0^\tau i(t)dt \approx I_{ON}\tau \ , \tag{2.1}$$

where we define the switching time, τ, to be the time to charge or discharge C_{sw} halfway. During most of the charging cycle, V_{SD} of the PMOS transistor is large, and for most of the discharging cycle, V_{DS} of the NMOS transistor is large, so these transistors are in the saturation region. It is reasonable, therefore, to assume the current is approximately constant during the charging and discharging cycles; $I_D \approx I_{DSAT}$, which we will call I_{ON}.

From Eq. (2.1), we find the switching time to be

$$\tau = \frac{C_{sw}V_{DD}}{2I_{ON}}.$$ (2.2)

Equation (2.2) illustrates an important point; the switching speed of a CMOS gate is determined by the DC saturation current of the transistors. The higher the current, the faster the switching capacitance can be charged and discharged. You might wonder why the time it takes for charge carriers to go from the source to the drain does not enter in. It does in principle and can be important for very high frequency RF transistors, but for digital circuits, charging and discharging times limit the speed.

The next question is: "What is the power dissipation of the CMOS gate?" If the gates are not switching, we have seen that only small leakage currents flow. This leads to a *static power* or *standby power* of

$$P_{standby} = N_G V_{DD} I_{OFF},$$ (2.3)

where N_G is the number of static gates, and I_{OFF} is the leakage current.

As shown in Fig. 2.4, when the input of a CMOS gate is driven by a clocked signal at a frequency, f, then the switching capacitance is continually being charged and discharged, and there is *dynamic power* dissipation. On the input low part of the cycle, the capacitor is charged through the PMOS transistor, and the result is that an energy

$$E_s = \frac{1}{2}C_{sw}V_{DD}^2$$ (2.4)

is stored on the capacitor.

How much energy is dissipated to charge the capacitor? While charging the capacitor, a current, $i(t)$, flows out of the power supply, so the power supply is delivering a power of $P(t) = i(t)V_{DD}$ to the circuit. Power is the derivative of energy, so the energy delivered by the power supply is

$$\begin{aligned} E &= \int_0^\infty i(t)V_{DD}\, dt = \int_0^\infty V_{DD}\frac{dq}{dt} dt \\ &= \int_0^{C_{sw}V_{DD}} V_{DD} dq = C_{sw}V_{DD}^2 \end{aligned},$$ (2.5)

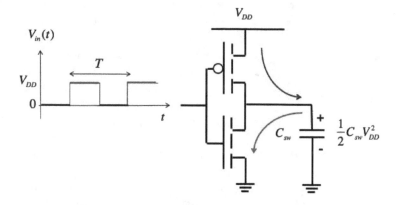

Fig. 2.4   The charging and discharging of the switching capacitance, C_{sw} that occurs when the input voltage cycles between the 0 and 1 states.

which is twice the energy stored in the capacitor. An energy of $C_{sw}V_{DD}^2/2$ has been dissipated in the PMOS transistor to charge the capacitor.

On the input high part of the cycle, the capacitor is discharged through the NMOS transistor, and an energy of $C_{sw}V_{DD}^2/2$ is dissipated in the NMOS transistor. When the gate is clocked at a frequency of $f = 1/T$, the dynamic power dissipated is

$$P = \frac{C_{sw}V_{DD}^2/2}{T/2} = fC_{sw}V_{DD}^2 .$$

Finally, we write the dynamic power dissipated by the switching gate as

$$P_{dynamic} = \alpha f C_{sw}V_{DD}^2 , \qquad (2.6)$$

where α, the *activity factor*, is the fraction of the time that the gate is switching, which is typically a few percent.

Equation (2.6) is an important result. The faster we run the circuit, the more power is dissipated. To run a processor fast and not dissipate excessive power, we need to make C_{sw} and V_{DD} as small as possible. Innovations such as *low-k dielectrics* have helped engineers minimize C_{sw}. Because the dynamic power is proportional to V_{DD}^2, reducing the power supply voltage can greatly reduce power dissipation, but it is important to maintain I_{ON} so that the switching speed does not suffer. Over several generations of integrated circuit technologies, V_{DD} was reduced from several volts to a little less than one volt, but as we'll learn later, the physics of MOSFETs makes it difficult to reduce the voltage much below one volt.

Figure 2.5 summarizes the design challenge. We start with a power budget. Too much power dissipation might overheat the chip or it may, just consume too much energy. As the number of transistor per chip increases, the capability of the electronic system increases, but even if the transistors are not switching, they still dissipate standby power. As the level of integration increases, the standby power increases, and there is less and less of the power budget available for the dynamic power that does the useful work of digital systems. A trade-off is necessary; in the most advanced, high-density digital systems, the power budget is divided roughly equally between standby and dynamic power.

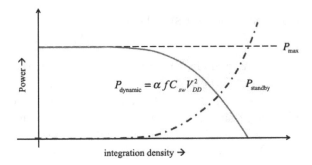

Fig. 2.5 The power-constrained design challenge. The sum of the standby and dynamic powers must not exceed the total power budget.

2.3 Analog/RF circuits

Our brief discussion of analog circuits will focus on the small signal amplifier shown in Fig. 2.6, the input voltage consists of a DC voltage, V_{IN} and a small AC signal, v_{in}, at the frequency, ω. The output current consists of a DC current, I_D, and a small AC signal, $i_d(\omega)$. The DC voltages and current are established by a bias circuit (not shown), which centers the operating point about which the AC signal moves in the saturation region. If the AC signal is small, then the problem can be solved by superposition, first the DC solution and then the AC solution. As shown on the right in Fig. 2.6, the DC power supply maintains a constant voltage, so it is an AC ground. To solve for v_{out} as a function of v_{in}, we need an AC small signal model for the transistor.

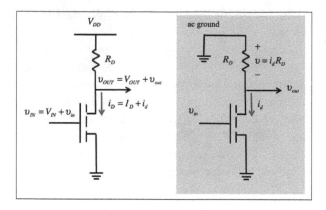

Fig. 2.6   The basic, common source (CS) analog amplifier. On the right is the AC, small signal circuit.

In the common source mode, the drain current is a function of v_{GS}, the total voltage, DC and AC, between the gate and the source, and v_{DS}, the total voltage between the drain and the source. The transistor characteristics we have been discussing describe the DC drain current in terms of the DC voltages. If we perturb these voltages a little, as the AC signal will do, then we can use a Taylor series to find the perturbed drain current,

$$I_D(V_{GS} + \delta V_{GS}, V_{DS} + \delta V_{DS}) = I_D(V_{GS}, V_{DS})$$
$$+ \left(\frac{\partial I_D}{\partial V_{GS}} \right)_{V_{DS}} \delta V_{GS} + \left(\frac{\partial I_D}{\partial V_{DS}} \right)_{V_{GS}} \delta V_{DS} + ... \qquad (2.7)$$

If the perturbations are small, we can ignore the higher order terms. From Eq. (2.7), we find

$$\delta I_D = g_m \, \delta V_{GS} + \delta V_{DS}/r_o$$
$$i_d = g_m \, v_{gs} + v_{ds}/r_o, \qquad (2.8)$$

where in the second line, we have associated the perturbed voltages with the small signal quantities. The assumption here is that the device can instantly follow the time-varying perturbed voltages. This *quasi-static assumption* is generally valid to quite high frequencies. Two important analog circuit parameters have been defined; the *transconductance*

$$g_m \equiv \left. \frac{\partial I_D}{\partial V_{GS}} \right|_{V_{DS}}, \qquad (2.9)$$

and the *output resistance*,

$$1/r_o \equiv \left. \frac{\partial I_D}{\partial V_{DS}} \right|_{V_{GS}}. \qquad (2.10)$$

The low-frequency small-signal equivalent circuit of the MOSFET shown at the top in Fig. 2.7 is a pictorial representation of Eq. (2.8). If the transistor on the right in Fig. 2.6 is replaced by this small signal equivalent circuit, then we can analyze the circuit to find the voltage gain as [1]

$$A_{v_i} = \frac{v_{out}}{v_{in}} = -g_m \left(R_D \| r_o \right) . \tag{2.11}$$

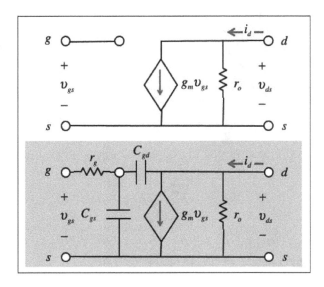

Fig. 2.7 Small signal equivalent circuit models for a MOSFET. Top: Low-frequency model. Bottom: High-frequency model. The gate resistance is not shown in the low-frequency model because the negligible DC gate current produces essentially no voltage drop.

In modern analog circuits, the resistor, R_D, is usually replaced by a transistor current source with high output resistance. The magnitude of the maximum gain that the transistor is capable of is

$$A_0 = g_m r_o , \tag{2.12}$$

which is known as the *self-gain*.

It is also important to know how high in frequency a transistor can operate. The high-frequency equivalent circuit of the MOSFET shown at the bottom of Fig. 2.7 shows that we must include the intrinsic capacitances of the transistor (as well as parasitic capacitances, which are not shown). One can see that as the frequency increases, the impedance of the capacitors will drop, which will reduce v_{gs} and lower the output current. If we AC short circuit the output and feed the transistor with a current source, $i_{in}(\omega)$, then we can analyze the circuit to determine $|\beta(\omega)| = |i_{out}/i_{in}|$ and plot $|\beta(\omega)|$ vs. frequency [1]. When $|\beta(\omega)| = 1$, we have reached a critical frequency beyond which the transistor does not amplify signals. This frequency is

$$\omega_T = 2\pi f_T = \frac{g_m}{C_{tot}}, \qquad (2.13)$$

where C_{tot} is the sum of the intrinsic and parasitic capacitances. The derivation is short (see p. 175 of [1], for example), and the answer is easy to remember. Transconductance is what amplifies signals; high transconductance is good, which is why it is in the numerator. Capacitance slows down circuits or shorts the signal out. Low capacitance is good, which is why it is in the denominator. The concept is very general. When a device engineer encountering a new device wants to estimate its high-frequency potential, she or he estimates the transconductance and total capacitance. The approach even works for digital circuits if we define a large signal transconductance

$$G_m = \frac{\Delta I_D}{\Delta V_{GS}} = \frac{I_{ON}}{V_{DD}}. \qquad (2.14)$$

This can be understood by recalling that if we set $V_{DS} = V_{DD}$, then for $V_{GS} = 0$, $I_D \approx 0$, and for $V_{GS} = V_{DD}$, $I_D = I_{ON}$. Recalling that $\omega = 1/\tau$, we can write by analogy with Eq. (2.13)

$$\frac{1}{\tau} = \frac{G_m}{C_{sw}} = \frac{I_{ON}}{C_{sw}V_{DD}}, \qquad (2.15)$$

which is essentially Eq. (2.2) for the switching speed of a digital gate. (The factor of 2 in Eq. (2.2) came because the switching time was defined to be the time to charge or discharge the capacitor halfway.)

2.4   MOSFET device metrics

Now that we understand a little about circuits, we can appreciate the importance of a few key device metrics. Figure 2.8 is a common source transfer characteristic with I_D plotted on a log axis so that I_{OFF}, which controls the standby power, can be seen. Also seen is I_{ON}, which controls the speed of a digital gate. It is important to note that below threshold, I_D varies exponentially with V_{GS} and above threshold, I_D varies as $(V_{GS} - V_T)^\alpha$, where $1 \leq \alpha \leq 2$.

The *subthreshold swing* is another important device metric:

$$SS \equiv \left(\frac{\partial(\log_{10} I_D)}{\partial V_{GS}} \right)^{-1}. \tag{2.16}$$

The subthreshold swing is always measured in the subthreshold region. As shown in Fig. 2.8, SS is the change in gate voltage needed to change I_D by a factor of 10; it is typically quoted in mV/decade. For a good transistor, SS is less than 100 mV/V.

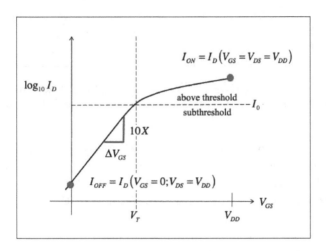

Fig. 2.8   A common source transfer characteristic, $\log I_D$ vs. V_{GS}, illustrating what is meant by the on-current and off-current. Also shown is the definition of the subthreshold swing (SS). An arbitrary small current, I_0, separates above threshold, $I_D > I_0$ from below threshold, $I_D < I_0$. The threshold voltage is V_{GS} at $I_D = I_0$.

Figure 2.9 illustrates why SS is so important. The two transfer characteristics with SS_1 and SS_2 have the same on-current, and therefore the same circuit speed, but the one with a higher SS has an exponentially larger off-current, which would result in much higher standby power. Device engineers work hard to achieve the smallest SS possible. As we will learn later, there is a lower limit to the SS of a MOSFET; it is 60 mV/decade at room temperature.

Figure 2.9 also illustrates another device metric, DIBL (Drain-Induced Barrier Lowering),

$$DIBL \equiv \left. \frac{\partial V_{GS}}{\partial V_{DS}} \right|_{I_D}. \tag{2.17}$$

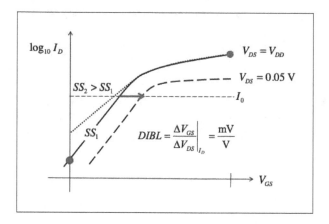

Fig. 2.9   A common source transfer characteristics illustrating how DIBL is measured. The dotted line shows why it is important to have a low subthreshold swing. A high SS increases the off-current exponentially.

If V_{DS} is reduced, the transfer characteristic shifts to the right. If V_{DS} is increased, the transfer characteristic shifts to the left. This shift can be interpreted as a change in threshold voltage with drain voltage, an undesirable effect for a transistor. As can be seen from Fig. 2.9, an increase in drain voltage lowers the threshold voltage. As shown in Fig. 2.9, DIBL is the change in gate voltage needed to maintain a constant subthreshold current when the drain voltage changes. It is usually quoted in mV/V. For a good transistor, $DIBL$ is typically less than 100 mV/V.

Figure 2.10 shows two different transfer characteristics for transistors with two different V_T's. (We define V_T to be the gate voltage needed to reach the current, I_0, which is at the boundary of the subthreshold and above threshold regions.) Characteristic (2) has a smaller V_T than characteristic (1). Consider what happens if we operate both transistors at V_{DD1}; characteristic (2) will have a larger on-current and therefore will be capable of operating at higher frequencies, but notice that this comes at a cost — the off-current and therefore the standby power of transistor (2) is exponentially higher than that of transistor (1). Alternatively, let's say that we reduce V_{DD} to V_{DD2}, which maintains the on-current. The speed of transistor (2) is the same as for transistor (1) because the on-currents are the same, but dynamic power goes as V_{DD}^2, so the dynamic power will be greatly reduced. Notice, however, that this also comes at a cost — the off-current and therefore the standby power of transistor (2) is exponentially higher that that of transistor (1), so even though V_{DD} is lower, the standby power has increased. We have neglected the effect of DIBL when lowering V_{DD} from V_{DD1} to V_{DD2}, but the general point is that there is a trade-off between on-current and off-current that comes from the characteristic shape of the MOSFET transfer characteristics. We will discuss the physical reasons for this characteristic shape later in these notes.

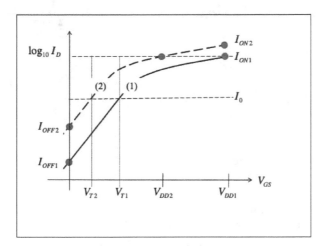

Fig. 2.10   Illustration of how reducing the threshold voltage from V_{T1} (solid line) to $V_{T2} < V_{T1}$ (dashed line) affects the on-current and the off-current.

Turning now to analog device metrics, three important metrics for analog applications are the transconductance, g_m, the output resistance, r_o, and their product, the self-gain, A_0. As shown in Fig. 2.11, the transconductance and output resistance are easily estimated from the output characteristics of the transistor. As shown on the left, we're interested in the transconductance in the saturation region where the DC bias places the transistor for analog applications. It is simply estimated from the measured I_D at adjacent gate voltages in the family of curves. As shown on the right, r_o is one over the slope of I_D vs. V_{DS} in the saturation region.

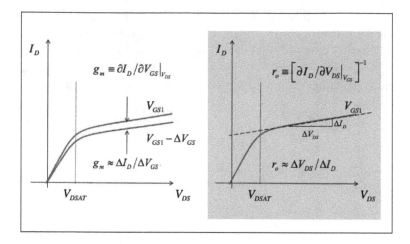

Fig. 2.11   Two important analog device metrics, small signal transconductance (left) and output resistance (right).

Most analog circuit design today is done with manufacturing processes designed for digital electronics. For digital electronics, the shortest possible channel lengths are used so that as many transistors as possible can be fit on the chip. We will see later, however, that the output resistance (and, therefore the self-gain) decreases as the channel length decreases, so analog designers use non-minimum length transistors.

While g_m and r_o are two of the most important analog metrics, several others are also important, such as device linearity, device noise, and device mismatch. Analog circuits often make use of source coupled pairs with a differential signal input applied between the gates. It's important that the two transistors be as nearly identical as possible. The maximum frequency of the circuit is limited by f_T and also by f_{max}, the maximum frequency of oscillation or the unity power gain frequency. The parameters, f_T and f_{max}, are differently affected by parasitics, so both are typically quoted to give designers a better understanding of how the transistor will perform in high-frequency circuits.

2.5 Summary

This lecture was a short introduction to digital and analog circuits. Several terms that will appear frequently in the remaining lectures have been introduced.

- CMOS inverter
- Pull up (pull down) transistor
- Noise margins
- Switching capacitance, C_{sw}
- On-current, I_{ON}
- Off-current, I_{OFF}
- Switching energy, $E_s = C_{sw}V_{DD}^2/2$
- Dynamic power, $P_{dynamic} = \alpha f C_{sw} V_{DD}^2$
- Standby (static) power, $P_{standby} = N_G V_{DD} I_{OFF}$
- Transconductance, $g_m = \partial I_D / \partial V_{GS}|_{V_{DS}}$
- Output resistance, $r_o = \left(\partial I_D / \partial V_{DS}|_{V_{GS}}\right)^{-1}$
- Self-gain, $A_0 = g_m r_o$
- Cutoff frequency, $f_T = g_m / C_{tot}$
- Subthreshold swing (SS)
- Drain-induced barrier lowering (DIBL)

So far, we have described the *IV* characteristics of transistors, but we have not explained them. Energy band diagrams provide us with a qualitative understanding of most of the features that we have described. The energy band approach to MOSFETs is the subject of the next lecture.

Lecture 2 Exercise: Device metrics

Shown in the figure below are the measured characteristics of n- and p-channel FinFETs. A good deal of information can be gleaned from measured IV characteristics. Reading from the graphs as carefully as possible, estimate some of device metrics.

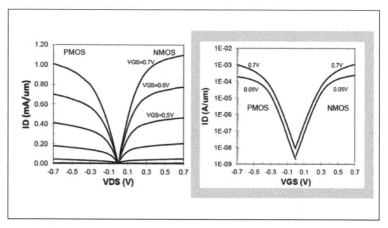

Measured IV characteristics of 14 nm FinFETs. (©IEEE 2014. Reprinted, with permission, from S. Natarajan, et al., "A 14nm Logic Technology Featuring 2nd-Generation FinFET Transistors, Air-Gapped Interconnects, Self-Aligned Double Patterning and a 0.0588 μm^2 SRAM cell size," pp. 70-72. Tech. Digest, Intern. Electron Dev. Mtg, Dec. 2014.)

1) The ON-current of the PMOS transistor.

From the output characteristics labeled PMOS, we see from the top curve for $V_{GS} = -0.7$ V that $I_{ON} \approx 1.0$ mA/μm. For the actual ON-current in mA, we would multiply this value by the width of the transistor in micrometers.

2) The OFF-current of the NMOS transistor.

The OFF-current is the current for $V_{GS} = 0$ and $V_{DS} = 0.7$ V. From the transfer characteristic labeled NMOS, we find $I_{OFF} \approx 10^{-8}$ A/μm or 10 nA/μm. For the actual OFF-current in mA, we would multiply this value by the width of the transistor in micrometers.

3) The DIBL of the PMOS transistor.
We measure $DIBL$ in the part of the transfer characteristic where I_D varies exponentially with V_{GS}. At $I_D = 10^{-7}$ A/μm, the difference in gate voltages for the two PMOS transfer characteristics is $\Delta V_{GS} \approx 50$ mV. The difference in drain voltages for these two curves is $\Delta V_{DS} = 0.65$ V, so $DIBL = \Delta V_{GS}/\Delta V_{DS} \approx 77$ mV/V.

4) The SS of the NMOS transistor.
We measure the SS swing at $V_{DS} = 0.7$ V on the NMOS transfer characteristic because, as we will discuss in the next lecture, high drain voltage can increase SS, so this is the worst case. At $I_D = 10^{-7}$ A/μm, $V_{GS} \approx 0.075$ V, and at $I_D = 10^{-6}$ A/μm, $V_{GS} \approx 0.15$ V, so $SS \approx 75$ mV/decade.

5) The V_{DSAT} of the NMOS transistor at $V_{GS} = 0.7$ V.
We can estimate V_{DSAT} as follows. For the $V_{GS} = 0.7$ V line in the NMOS output characteristic, draw a straight line beginning at $V_{DS} = 0$ that is tangent to the linear region and another one beginning at $V_{DS} = 0.7$ V that is tangent to the saturation region. These two lines intersect at $V_{DS} \approx 0.2$ V, which we can take as an estimate of V_{DSAT}; $V_{DSAT} \approx 0.2$ V.

6) The threshold voltage, V_T, of the NMOS transistor.
We can only roughly estimate V_T from this data. Note that the output characteristics for the NMOS transistor show no current at $V_{GS} = 0.2$ V and a small current at $V_{GS} = 0.3$ V, so $V_T \approx 0.3$ V. Note from the logarithmic plot of I_D in the transfer characteristics that nothing special happens at $V_{GS} = V_T$. The drain current increases monotonically from $V_{GS} = 0$ to 0.7 V; V_T is just a voltage at which "significant" current begins to flow.

7) The NMOS transistor linear region resistance at $V_{GS} = 0.7$ V.
For the $V_{GS} = 0.7$ V line in the NMOS output characteristic, draw a straight line beginning at $V_{DS} = 0$ that is tangent to the linear region. This line reaches 1.20 mA/μm at $V_{GS} \approx 0.25$ V, so $R_{tot} = 0.25$ V/1.20 mA/μm ≈ 0.2 k$\Omega - \mu$m. For the actual resistance in Ohms, we would divide this value by the width in micrometers. We label this as the total resistance because it includes the channel resistance plus series resistances at the source and drain contacts.

8) The g_m of the PMOS transistor at $V_{GS} = -0.7$ V.

To estimate g_m from $\Delta I_D / \Delta V_{GS}$, we should have closely spaced gate voltages. The gate voltages in the output characteristics are spaced by 0.1 V, but g_m is fairly constant, so we should be able to get a good estimate. From the PMOS output characteristic, we find $I_D(V_{GS} = -0.7, V_{DS} = -0.7) \approx 1.0$ mA/μm, and $I_D(V_{GS} = -0.6, V_{DS} = -0.7) \approx 0.7$ mA/μm, so $\Delta I_D / \Delta V_{GS} \approx (0.3$ mA/μm$)$ / 0.1 V $= 3$ mS/μm; $g_m \approx 3$ mS/μm.

9) The r_o of the PMOS transistor at $V_{GS} = -0.7$ V.

In the PMOS output characteristic for the top curve with $V_{GS} = -0.7$ V, draw a straight line tangent to the curve beginning at $I_D = 1.0$ mA/μm. This line reaches $I_D = 0.80$ mA/μm at $V_{DS} = 0.05$ V. From $r_0 = \Delta V_{DS} / \Delta I_D = 0.65$ V / 0.2 mA/μm, so $r_o \approx 3.25$ k$\Omega - \mu$m. For the actual resistance in Ohms, we would divide this value by the width in micrometers.

10) The self-gain, A_0 of the PMOS transistor at $V_{GS} = -0.7$ V.

The self-gain is $A_0 = g_m r_o \approx 3 \times 3.25$, so $A_0 \approx 9.75$.

2.6   References

The basics of digital and analog circuits are covered in almost any introductory electronics book. A good place to begin is with Chapters 7 and 14 of Sedra and Smith.

[1] Adel S. Sedra and Kenneth C. Smith, *Microelectronic Circuits*, Seventh Ed., Oxford Univ. Press, New York, 2015.

Lecture 3

IV Theory: Energy band approach

3.1 Introduction

Most transistors operate by controlling the height of an energy barrier with an applied voltage. Examples are field-effect transistors (FET's), such as MOSFETs, JFETs (junction FET's), and HEMTs (high electron mobility transistors) as well as bipolar junction transistors (BJTs) and heterojunction bipolar transistors (HBTs) [1, 2]. The operating principles of these transistors are most readily understood in terms of energy band diagrams, a powerful approach for qualitative understanding. In the rest of these lectures, we will develop mathematical models for the IV characteristics, but before we do that, we need the clear physical understanding that energy band diagrams provide.

3.2   Review of energy band diagram

The operation of a semiconductor device is governed by the Poisson equation and the electron and hole continuity equations [3]. The solution to this rather complicated set of coupled, nonlinear partial differential equations gives the electrostatic potential, $\psi(\vec{r})$, and electron and hole carrier concentrations, $n(\vec{r})$ and $p(\vec{r})$, as a function of position. Drawing these energy band diagrams takes a little intuition about the solution to the semiconductor equations, but a little intuition goes a long way. Energy band diagrams are a powerful conceptual tool in the device engineer's toolkit. Essentially all of the features of transistor characteristics discussed in the previous two lectures can be understood with energy band diagrams.

An energy band diagram is a plot of the bottom of the conduction band and top of the valence band versus position. Figure 3.1 is an example that illustrates the basic principles. On the upper left is an equilibrium energy band diagram for a uniformly doped n-type semiconductor with the electron density equal to the ionized donor density, $n_0 = N_D^+$. Recall that n_0 is related to the relative position of the Fermi level with respect to the band edge according to

$$n_0 = N_C e^{(E_F - E_c)/k_B T}, \tag{3.1}$$

where N_c is the effective density-of-states of the conduction band. Mobile electrons distribute themselves to cancel the charge of the ionized donors. The result is that this semiconductor is charge-neutral.

The lower left in Fig. 3.1 illustrates what happens when we apply a bias across the semiconductor. Because the semiconductor is charge-neutral, the electric field in the semiconductor is constant. Recalling that the electric field is minus the gradient of the electrostatic potential, $\mathcal{E} = -d\psi/dx$, we see that the electric, $\mathcal{E} = -V_A/L$, points from right to left. The electric field exerts a force on electrons, $F_e = -q\mathcal{E}$ that is opposite in sign to the electric field (q is the magnitude of the charge on an electron). The force on electrons points from the left to the right in Fig. 3.1. Imagine pushing an electron from $x = L$ to $x = 0$; we need to do work against the force and push the electron "uphill," which raises its potential energy. We conclude that the electron potential energy at $x = 0$ is higher than at $x = L$. How much lower is the electron energy at $x = L$ than at $x = 0$?

Recall that force is minus the gradient of energy, U, or

$$\frac{dU}{dx} = -F_e = q\frac{d\psi}{dx}, \tag{3.2}$$

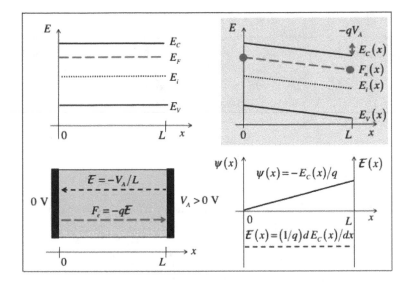

Fig. 3.1 Illustration of energy band diagram fundamentals. Upper left: Equilibrium energy band diagram of an n-type semiconductor. Lower left: Illustration of the electric field, $\mathcal{E}$, and force on electrons, F_e for the same semiconductor with an applied bias. Upper right: The energy band diagram under bias. Lower right: The electrostatic potential and electric field vs. position.

which can be integrated across the semiconductor to find

$$\int_{U(0)}^{U(L)} dU = \int_0^L -F_e \, dx = -\int_0^L q \left(\frac{d\psi}{dx}\right) dx \tag{3.3}$$

$$U(L) - U(0) = q\left(\psi(0) - \psi(L)\right) = q\left(V(0) - qV(L)\right) = -qV_A.$$

We use $\psi(x)$ to denote the electrostatic potential in the semiconductor and V to denote voltages on the contacts. We conclude that the applied voltage, V_A, at the right contact lowers the electron potential energy there by qV_A. Similarly, at any position, x, the electrostatic potential lowers the electron potential energy by $q\psi(x)$. The bottom of the conduction band is the potential energy of electrons in the conduction band, so

$$E_c(x) = E_{co} - q\psi(x), \tag{3.4}$$

where E_{co} is E_c where $\psi(x) = 0$. A positive electrostatic potential in the semiconductor lowers E_c. It also lowers $E_v = E_c - E_g$, where E_g is the band gap of the semiconductor.

As shown on the upper right of Fig. 3.1, we can now draw the energy band diagram of the semiconductor under bias. The equilibrium Fermi level is replaced by the electron quasi-Fermi level, F_n, the analogous quantity out of equilibrium. (In general, there is a hole quasi-Fermi level too, but in this example, the two are identical, $F_p = F_n$.) The ideal ohmic contacts assumed connect to F_n. At $x = 0$, the potential is the same as in equilibrium, so $F_n(x = 0)$ is where the Fermi level was in equilibrium. At $x = L$, F_n is lowered by qV_A. A positive voltage applied to a contact lowers the quasi-Fermi level there.

In general, $F_n(x)$ inside the device is determined by a solution to the semiconductor equations, but we can use our intuition. Recall that the electron current is proportional to the slope of F_n, $J_n = n\mu_n \nabla F_n$ (p. 205 in [3]). Electrons flow through the semiconductor and around the external circuit providing the bias. We expect to find $n(x) = N_D^+$, just as in equilibrium, because charge neutrality will be maintained. The electric field and the electron density will be constant, so the current will be constant and, therefore, so will the slope of $F_n(x)$, as shown in the figure.

We have deduced how F_n varies with position; what about $E_c(x)$? Out of equilibrium, Eq. (3.1) becomes

$$n(x) = N_c e^{(F_n(x) - E_c(x))/k_B T} . \tag{3.5}$$

Since $n(x) = n_0$ to maintain space-charge neutrality, $E_c(x)$ must follow $F_n(x)$ to maintain the constant electron density. The result of this reasoning is the energy band diagram on the upper right.

Given an energy band diagram like that on the upper right of Fig. 3.1, we can "read" it to extract useful information. For example, by taking the gradient of $E_c(x)$, we find from Eq. (3.4),

$$\mathcal{E} = \frac{1}{q} \frac{dE_c}{dx} ; \tag{3.6}$$

the electric field is proportional to the slope of the band edge, as shown on the lower right in Fig. 3.1. We also see from Eq. (3.4) that when $\psi(x)$ increases, $E_c(x)$ decreases, so we can sketch $\psi(x)$ vs. x by flipping $E_c(x)$ or $E_v(x)$ upside down, as also shown on the lower right.

With this short review of energy band diagrams, we are ready to discuss an energy band treatment of the MOSFET.

3.3 Equilibrium

A full understanding of MOSFETs requires multi-dimensional energy band diagrams, but the essentials can be conveyed in 1D. In this section, we examine $E_c(x)$ along the surface of a bulk (or planar) MOSFET as indicated in Fig 3.2. We'll show that the *IV* characteristics discussed in the first two lectures can be explained by understanding how $E_c(x)$ changes when voltages are applied to the contacts. Later, in Lecture 5, we will examine energy band diagrams normal to the surface, in the *y*-direction.

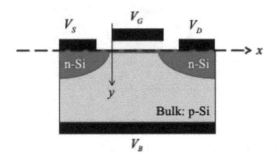

Fig. 3.2   The cross section of a bulk (or planar) MOSFET showing the line along the Si surface for which we will sketch the energy vs. position, $E_c(x, y = 0)$, from the source, across the channel, to the drain. The *z*-axis is into the page, in the direction of the width of the transistor, W.

The source and drain of an n-channel MOSFET are heavily doped n-type, and the channel of a bulk MOSFET like that in Fig. 3.2 is p-type. In equilibrium, the Fermi level is constant as shown in Fig. 3.3. In the n-type source and drain regions, we draw $E_c(x)$ near E_F, and in the p-type channel we draw $E_v(x)$ near E_F. After smoothly connecting the bands, we find the equilibrium energy band diagram as shown in Fig. 3.3. Note the potential energy barrier that separates electrons in the source from electrons in the drain. This barrier plays a critical role in the operation of transistors. Because the bands vary with position, we know that there is a position-dependent electrostatic potential in the semiconductor, and $E_c(x)$ varies according to Eq. (3.4). Similarly, $E_v(x) = E_c(x) - E_g$ also tracks $E_c(x)$. Because E_c is lower in the source and drain than in the channel, the electrostatic potential in the source and drain is higher than in the channel.

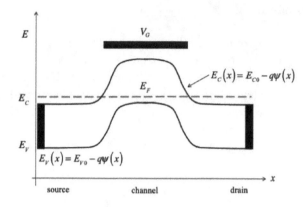

Fig. 3.3 The equilibrium energy band diagram along the top surface of an n-channel MOSFET. Also shown schematically is the gate electrode. A voltage on the gate affects the electrostatic potential in the channel. The Fermi level is above E_c in the source and drain because these regions are degenerately doped.

The next step is to understand how the energy bands change when voltages are applied to the gate and drain terminals (the source terminal is grounded). Recall that we use $\psi(x)$ to denote the electrostatic potential in the semiconductor and V to denote voltages on the contacts. When we apply a voltage, V_{DS}, between the drain and the source, the Fermi level is replaced by a position-dependent electron quasi-Fermi level, and current will flow (see [3], p. 205). When we apply a voltage, V_G, to the gate, the electrostatic potential in the semiconductor channel will be affected because the gate electrode is very close to the channel — separated only by a very thin oxide.

3.4   Application of a gate voltage

Figure 3.4 shows what happens when a positive voltage is applied to the gate. In this figure, we show only the conduction band, because we are discussing an n-channel MOSFET for which the current is carried by electrons in the conduction band. As shown on the left, for $V_S = V_D = 0$, the Fermi levels in the source, channel, and drain all align; the device is in equilibrium, and no current flows. The application of a gate voltage does not disturb equilibrium because the gate electrode is separated from the silicon by a thin gate oxide insulator. A positive gate voltage does, however, increase the electrostatic potential in the channel, which lowers the conduction band according to Eq. (3.4)

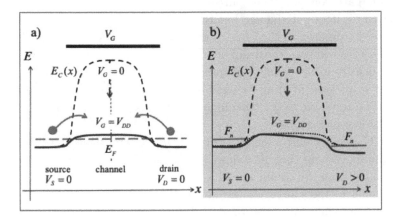

Fig. 3.4   Sketch of $E_c(x)$ for an n-channel MOSFET. Left a): The voltages on the source and drain are zero. Right b): A small positive voltage is applied to the drain. In both cases, two gate voltages are shown, a low gate voltage, $V_G = 0$, (dashed line) and a high gate voltage, $V_G = V_{DD}$, (solid line).

For low gate voltage (the dashed line for $V_G = 0$), the energy barrier between the source and drain is high, which prevents electrons from entering the channel. Under high gate voltage (the solid line for $V_G = V_{DD} > V_T$), the energy barrier is low, and the electrons' thermal energy of $k_B T$ allows electrons from the source and drain to surmount the barrier and enter the channel. As a result, when $V_G > V_T$, there is a high density of electrons in the channel. The channel, with its uniform concentration of electrons is like the n-type semiconductor in equilibrium shown on the upper left in Fig. 3.1 except that the high electron concentration is not produced by n-type doping but, rather, by the positive voltage on the gate.

It is important to note that the application of a gate voltage does not affect the Fermi level in the underlying silicon. The Fermi level in the device can only change if the source or drain voltages change, because the source and drain contacts are connected to the Fermi level in the device, and the gate electrode is not. We conclude that the application of a gate voltage simply raises or lowers the potential energy barrier between the source and drain. As shown in Fig. 3.4a, the device remains in equilibrium, and no current flows.

3.5   Application of a small drain voltage

Figure 3.4b shows what happens when we add a small drain voltage to the large gate voltage. The source is grounded, so the quasi-Fermi level in the source does not change, but the small, positive drain voltage lowers the quasi-Fermi level in the drain. Lowering F_n in the drain lowers E_c too. Electrostatics will keep the drain neutral, so $n \approx n_0 \approx N_D$, where N_D is the doping density in the drain. The quantity, $F_n - E_c$ determines the electron density, so to keep $n(x)$ at its equilibrium value, $E_c(x)$ must drop in the drain. The resulting $E_c(x)$ for a high V_G and a small $V_{DS} > 0$ is shown as the solid line in Fig. 3.4b.

For $V_{GS} > V_T$ and a small, positive V_{DS}, electrons enter the channel from the source and then flow down hill to the drain. Electrons also enter the channel from the drain, but it is harder for them to flow uphill across the channel to the source, so the net flow of electrons is from the source to the drain, which gives rise to a small drain current. As the small drain voltage increases, the net flow of electrons from the source to the drain increases. This is the linear region of the MOSFET *IV* characteristic.

Under small drain bias, the channel behaves very much like the n-type resistor shown in Fig. 3.1. We see from the slope of $E_c(x)$ in Fig. 3.4b that the electric field is constant in the channel, just as it was in the resistor of Fig. 3.1. The electron density is also constant in the channel, so $F_n(x)$ varies linearly in the channel, just as it did in the resistor of Fig. 3.1. If we were to increase the gate voltage further, there would be more electrons in the channel and its resistance would drop. Under small V_{DS}, the MOSFET behaves as a gate voltage controlled resistor.

3.6   Application of a large drain voltage

Figure 3.5 shows what happens when a large drain voltage is applied. The source is grounded, so the quasi-Fermi level in the source does not change, but the positive drain voltage lowers the quasi-Fermi level in the drain. Lowering the Fermi level lowers E_c too, to keep the drain neutral. The resulting energy band diagrams are shown in Fig. 3.5. Note that we have only shown the quasi-Fermi levels in the source and drain, but $F_n(x)$ varies smoothly across the device. Numerical simulations are needed to resolve $F_n(x)$ across the channel, but it is clear that there will be a slope to $F_n(x)$, so current will flow.

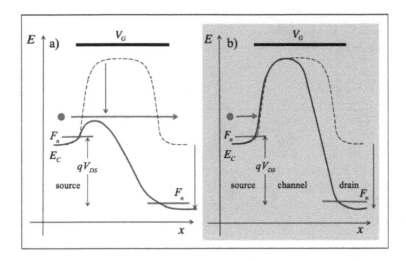

Fig. 3.5 Sketch of $E_c(x)$ for an n-channel MOSFET. The voltage on the source is zero, and the voltage on the drain is large. Left a): High gate voltage ($V_{GS} > V_T$) and high drain voltage ($V_{DS} > V_{DSAT}$). Right b): Low gate voltage ($V_{GS} = 0$) and high drain voltage. The figure on the left is for the on-state, and the figure on the right is for the off-state. In both cases, the dashed line shows the corresponding $V_{GS} = V_{DS} = 0$ equilibrium energy band diagram.

Consider first the on-state shown in Fig. 3.5a. The current is controlled by the height of the energy barrier between the source and channel. In an electrostatically well-designed MOSFET, the height of this energy barrier is strongly controlled by the gate voltage and only weakly controlled by the drain voltage. If the gate voltage increases, the barrier lowers, and more electrons in the source can surmount the barrier and flow across the channel to the drain. Electrons from the drain can also enter the channel, but they see a large energy barrier to the source, so they cannot flow across the channel to the source as some could under low drain bias. Under high drain bias, electron flow is essentially in one direction, from the source to the drain. If the drain voltage increases, it has little effect on the energy barrier at the source, so there is little effect on the drain current. This is the saturation region of the MOSFET *IV* characteristic. When the gate and drain voltages are high, the transistor is in the *on-state* with the current being the on-current, $I_{ON} = I_D(V_{GS} = V_{DS} = V_{DD})$ of Fig. 2.8.

Next, consider what happens when we keep the drain voltage high but reduce the gate voltage to $V_G = 0$. The resulting energy band diagram is shown in Fig. 3.5b. Note that even though V_{DS} is large, very little current

flows because there is a very small probability that electrons in the source
can surmount the large source to channel energy barrier. The transistor is
in the *off-state* with the current being the off-current, $I_{OFF} = I_D(V_{GS} = 0, V_{DS} = V_{DD})$ of Fig. 2.8.

3.7   Transistor operation

Figure 3.6 summarizes how modulating the height of an energy barrier with
a gate voltage produces the *IV* characteristics of a MOSFET. Shown on the
right are the measured output characteristics of an $L = 105$ nm n-channel
MOSFET. Shown on the left are numerical simulations of $E_c(x)$ at various
gate voltages in the linear and saturation regions of operation. Note that
under high gate voltage in the linear region, $E_c(x)$ varies linearly with x,
which corresponds to a constant electric field, as expected in the linear
region where the MOSFET acts as a gate voltage controlled resistor.

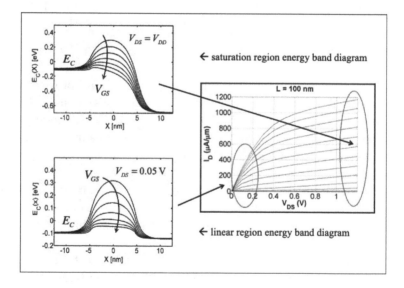

Fig. 3.6   Illustration of how the common source output characteristics of an n-channel
MOSFET are explained with energy band diagrams. On the right are the measured *IV*
characteristics of an $L = 105$ nm MOSFET (data provided by Shuji Ikeda, ATDF, Dec.
2007). On the left are numerical simulations of $E_c(x)$ vs. x for a different MOSFET.
(©IEEE 2002. Reprinted, with permission, from: Mark Lundstrom and Zhibin Ren,
"Essential Physics of Carrier Transport in Nanoscale MOSFETs," *IEEE Trans. Electron
Dev.*, **49**, pp. 133-14, 2002.)

Figure 3.6 also shows simulations of $E_c(x)$ in the saturated region. As the gate voltage pushes the potential energy barrier down, electrons in the source hop over the barrier and then flow down hill to the drain. This figure also illustrates why the drain current saturates with increasing drain voltage. It is the barrier between the source and channel that limits the current. Electrons that make it over the barrier flow down hill and out the drain. Increasing the drain voltage (assuming that it does not lower the source to channel barrier) does not increase the current. Note also that even under very high gate voltage, a small barrier remains. Without this barrier and its modulation by the gate voltage, we would not have a transistor.

Figure 3.7a shows the measured transfer characteristics of the same $L = 105$ nm MOSFET shown in Fig. 3.6. Figure 3.7b shows the measured transfer characteristic of a shorter channel length device ($L = 85$ nm) in the same technology. The DIBL observed in these characteristics is a short channel effect that will be discussed in the next section. Let's focus now on the well-behaved I_D vs. V_{GS} characteristic for $V_{DS} = 1.2$ V shown in Fig. 3.7a. We observe that below threshold, I_D varies exponentially with V_{GS}, and above threshold, I_D varies more slowly with V_{GS} because the mobile charge "screens" the channel from the influence of the gate potential. The energy band diagram can help us understand why I_D varies exponentially with V_{GS} below threshold.

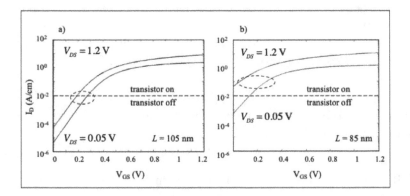

Fig. 3.7 Measured transfer characteristics of $L = 105$ nm and $L = 85$ nm MOSFETs. (Data provided by Shuji Ikeda, ATDF, Dec. 2007).

The off-current is small because the probability that an electron in the source can surmount the large off-state energy barrier shown in Fig. 3.5b is small. If E_b is the height of the energy barrier, then the probability that an electron in the source has this energy is given by the Fermi function as

$$f(E) = \frac{1}{1 + e^{(E_b - E_F)/k_B T}} \approx e^{-(E_b - E_F)/k_B T}, \qquad (3.7)$$

where the approximation comes from the fact that $E_b \gg E_F$ in the off-state. We conclude that in the off state

$$I_D \propto e^{-E_b/k_B T}. \qquad (3.8)$$

Two different barrier heights produce two different currents,

$$\frac{I_{D2}}{I_{D1}} = \frac{e^{-E_{b2}/k_B T}}{e^{-E_{b1}/k_B T}} = e^{\Delta E_b/k_B T}, \qquad (3.9)$$

where ΔE_b is the barrier height lowering. How much lower must E_b be to increase the current by a factor of 10? From Eq. (3.9) we find

$$\Delta E_b(eV) = (\ln 10)(k_B T/q) = 0.060 \text{ eV}. \qquad (3.10)$$

The subthreshold current of a MOSFET is a result of *thermionic emission* of electrons over an energy barrier; lowering the barrier by 60 meV at $T = 300$ K increases the current by a factor of 10 (a decade). A positive potential in the semiconductor lowers the barrier and increases the current, so from Eq. (3.8), we find

$$I_D \propto e^{-E_c(x=0)/k_B T} \propto e^{q\psi_s(x=0)/k_B T}, \qquad (3.11)$$

where $E_c(x = 0)$ is the bottom of the conduction band at the top of the barrier, and $\psi_s(x = 0)$ is the potential there. The gate voltage is what changes the potential in the semiconductor, but only a fraction of the gate voltage gets into the semiconductor. As we will see in Lecture 5,

$$\psi_s(x = 0) = \frac{V_{GS}}{m}, \qquad (3.12)$$

where $m \geq 1$ is a number. Finally, using this result in Eq. (3.11), we find for the subthreshold current

$$I_D \propto e^{q V_{GS}/m k_B T}. \qquad (3.13)$$

If we use Eq. (3.13) to find the change in V_{GS} needed to increase the current by a factor of 10 (i.e. the subthreshold swing), we find

$$SS = (\ln 10)m(k_B T/q) = 2.3m(k_B T/q). \qquad (3.14)$$

If $m = 1$, then a change in V_{GS} produces the same change in ψ_s, and the subthreshold swing has its minimum value of 60 mV/decade at room temperature. The value of m is determined by the electrostatic design of the MOSFET. Transistor designers work hard to make m as close to one as possible.

Figure 3.7a also shows that above threshold, I_D varies more slowly with V_{GS}. Equation (3.11) continues to apply above threshold, but Eq. (3.13) does not. As we will discuss in Lecture 5 on MOS electrostatics, above threshold it is difficult for the gate to push the energy barrier down. Instead of ψ_s being proportional to V_{GS} as in Eq. (3.12), we will find that above threshold, ψ_s is proportional to $\ln(V_{GS})$. This occurs because the high density of electrons at the top of the barrier screens out the gate potential.

3.8   Short channel effects

For many years, progress in microelectronics occurred by shrinking the channel length of MOSFETs to reduce their overall size and allow more to be placed on an integrated circuit chip. The challenge in downscaling the channel length was to deal with various *short channel effects* and preserve good electrical characteristics of the transistor. One short channel effect has to do with the output resistance, r_o, of the transistor. As shown in Fig. 3.6, I_D does not fully saturate in the so-called saturation region. The output resistance decreases as the channel length decreases. Another short channel effect is DIBL, as shown in Fig. 3.7a. As the channel length decreases, DIBL increases. Finally, it's also possible that the subthreshold swing can increase, as shown in Fig. 3.7b.

For each technology generation, there is a minimum channel length below which transistors have degraded performance. For the technology shown in Figs. 3.6 and 3.7, the $L = 105$ nm MOSFET is well behaved, but the $L = 85$ nm MOSFET is not. For the next technology generation, the transistors are re-designed to suppress short channel effects and produce transistors that are well-behaved at a shorter channel length. Although short channel effects are the result of some rather complex two- and three-dimensional electrostatics, the general principles are readily understood in terms of energy band diagrams.

Figure 3.8a illustrates the meaning of "drain-induced barrier lowering" (DIBL). In a well designed transistor, the energy barrier between the source and channel is completely controlled by the gate voltage. For very long channel MOSFETs, there is no DIBL, but for short channel MOSFETs, the positive voltage on the drain produces an electric field that can reach through to the source and lower the barrier as illustrated in Fig. 3.8a. The

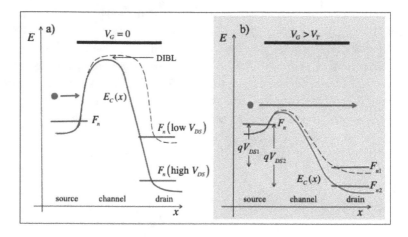

Fig. 3.8   Sketches of $E_c(x)$ that illustrate the physics of DIBL and output resistance. Left a): The off-state with low gate voltage ($V_{GS} = 0$) and two different drain voltages, a low V_{DS} (dashed line) and a high V_{DS} (solid line).   Right b): The on-state with high gate voltage ($V_{GS} > V_T$) and two different drain voltages, V_{DS1} (dashed line) and $V_{DS2} > V_{DS1}$ (solid line). (Both V_{DS1} and V_{DS2} are greater than V_{DSAT}.)

lower barrier means that at $V_{GS} = 0$, the off-current is increased. Also, at any specific $V_{GS} > 0$, the current is higher because the barrier is lower. The gate voltage needed to reach a specific I_D is lower because the drain helps the gate pull the barrier down. The threshold voltage is reduced because the V_{GS} needed to push the barrier down to its on-state height is smaller. These considerations explain the *IV* characteristics of a well-behaved short channel transistor like the one illustrated in Fig. 3.7a.

The *IV* characteristic in Fig. 3.7a displays DIBL, but the subthreshold swing is essentially the same for low and high V_{DS}. For the shorter channel length MOSFET in Fig. 3.7b, SS at low V_{DS} is larger, and SS at high V_{DS} is still larger. The off-current is so high that this transistor is unusable, and it is hard to define a clear DIBL number because the two subthreshold characteristics are not parallel. We can understand what is happening with the aid of Eq. (3.12). The gate and the drain share control of the height of the energy barrier, so a change in V_{GS} does not produce the same change in ψ_s at the top of the barrier. The parameter, m, in Eq. (3.12) is greater than one, so the subthreshold swing increases according to Eq. (3.14).

Figure 3.8b illustrates a similar effect of the drain voltage for $V_{GS} > V_T$. The energy barrier is small, so the drain current is high, but if the drain voltage increases, it can lower the barrier a little, so the current increases.

This explains why the drain current does not completely saturate in a short channel transistor, i.e. why the output resistance, r_o, is not infinite.

We expect DIBL to increase, V_T to decrease, and r_o to decrease as L decreases because a reduced channel length brings the drain closer to the source, but as sketched in Fig. 3.9, if the channel length is too short, the results can be catastrophic. The transfer characteristics sketched in Fig. 3.9a show that if the channel length is too short, the gate loses control over the barrier. Figure 3.9b shows what happens when the drain has significant control of the barrier height; there is no clear saturation region, and an output resistance cannot be defined. This device is said to be *punched through* — the drain potential has reached across the channel and pulled the barrier down so that current can flow with little help from the gate. We no longer have a good transistor.

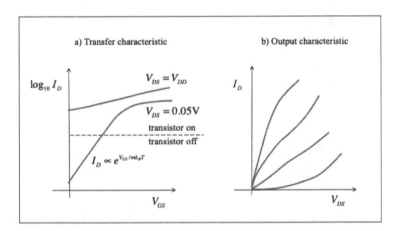

Fig. 3.9   Illustration of a MOSFET suffering from severe short channel effects. Left a): Transfer characteristics. Right b): Output characteristics.

3.9   Discussion

We have discussed how a transistor's *IV* characteristics can be qualitatively understood with energy band diagrams, and we have also used energy band diagrams to explain important short channel effects that occur as L is scaled down. The electrostatic design of a MOSFET to minimize short channel effects and maximize on-current at low voltage is the most challenging aspect of MOSFET design for digital electronics. Figure 3.10 illustrates the electrostatic challenge.

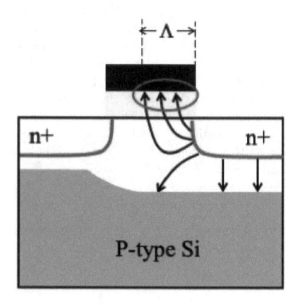

Fig. 3.10   Illustration of geometric screening in a bulk (or planar) MOSFET. The arrows represent the electric field lines caused by the positive voltage on the drain. Some field lines terminate on the metal gate and some in the p-type bulk. In a well-designed MOSFET, the distance that they penetrate, Λ, is less than L so that they do not reach through to the source and cause DIBL. (After David Frank, Yuan Taur, and Hon-Sum Philip Wong, "Future Prospects for Si CMOS Technology," Technical Digest, IEEE Device Research Conf., pp. 18-21,1999.)

From the demonstration of the first Si MOSFET in 1959 until the early 2000s, integrated circuit technology was largely bulk, planar, MOSFET technology. To preserve good electrical characteristics as L was scaled down, the gate oxide was thinned from 120 nm initially to just over 1 nm eventually. This put the gate electrode closer to the channel and increased its control of the potential at the top of the barrier. The channel doping was also increased to make it harder for the drain electric field to reach the source. Sophisticated channel doping profiles were developed (e.g. so-called halo implants), but to preserve good electrical characteristics at $L = 22$ nm and below, a new transistor structure was needed.

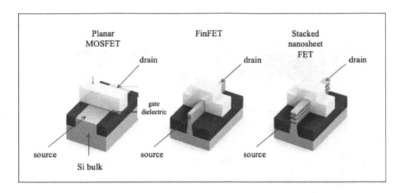

Fig. 3.11   Evolution of the MOSFET. Left: Bulk (planar) MOSFET. Middle: FinFET. Right: Stacked nanosheet MOSFET. (©IEEE 2019. Reprinted, with permission, from: Peide Ye, Thomas Ernst, and Mukesh V. Khare, "The Last Silicon Transistor" *IEEE Spectrum*, pp. 30-35, August, 2019.)

Figure 3.11 illustrates the evolution of the MOSFET. Beginning in 1959 with the demonstration of the first Si MOSFET and for the next 50 years, most MOSFETs had a planar geometry. Beginning in 2011 with the 22 nm technology node, adequate electrostatic control of short channel effects could not be achieved with the planar geometry. At that point, the industry shifted to the FinFET geometry shown in the middle of Fig. 3.11. In the FinFET, the channel is a vertical fin of silicon. Wrapping the gate around three sides of the fin gives the gate better control of top of the source to channel energy barrier. As channel lengths continue to scale down, eventually *gate all around* structures like the stacked nanosheet geometry shown at the right in Fig. 3.11 will be needed. In this structure, each Si nanosheet is encased in a thin gate dielectric covered by a metal gate electrode. By stacking nanosheets vertically, high on-currents can be achieved in a small transistor area. Such a device may be the ultimate field-effect transistor in terms of size and performance.

If the electrostatics can be controlled, is there a fundamental lower limit to the channel length? This question can be explored by numerical device simulations that treat electrons as quantum mechanical particles. Figure 3.12 shows simulations of quantum transport in a gate all around nanowire Si MOSFET. Electrostatic control for this model device is excellent, so the scaling limits are determined by quantum mechanical tunneling of electrons from the source through the source to channel barrier in the off-state. The plots show the energy-resolved current in the transistor.

Transistors!

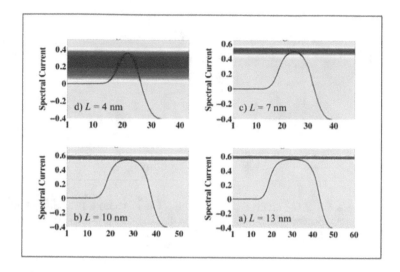

Fig. 3.12   The energy-revolved current as computed by a quantum transport simulation for a silicon nanowire MOSFET in the off-state under high drain bias. The nanowire is < 110 > oriented and is 3 nm in diameter. The x-axis is in arbitrary units. (Simulations performed by Dr. Mathieu Luisier, ETH Zurich and used with permission, 2014).

At $L = 13$ nm a) on the lower right, the off-state leakage current flows over the top of the barrier. This transistor is operating as a classical, barrier-controlled device. When the channel length decreases to 10 nm (b, lower left), a small fraction of the current begins to tunnel through the barrier. At 10 nm, however, good transistor performance is still obtained. For $L = 7$ nm (c, upper right), a substantial fraction of the off-state current flows by tunneling under the barrier. The performance of the device (e.g. its subthreshold slope) degrades. Finally, at $L = 4$ nm (d, upper left) most of the off-state current is due to tunneling through the barrier. At this channel length, it is difficult to modulate the current by controlling the barrier height because the barrier is transparent to electrons. When the barrier is too thin and becomes transparent to electrons, we no longer have a good transistor. Modern transistors are getting close to this fundamental limit.

3.10 Summary

Transistor physics boils down to electrostatics and transport. The energy band diagram is a qualitative illustration of transistor electrostatics. In practice, most of transistor design is about engineering the device so that the energy barrier is appropriately manipulated by the applied voltages. The design challenges have increased as transistors have gotten smaller and smaller, and we understand transistor electrostatics better now, but the basic principles are the same as they were in the 1960s.

In a well-designed MOSFET, the region near the top of the barrier is under the strong control of the gate voltage and only weakly affected by the drain voltage. The goal in transistor design is to achieve this performance as channel length scaling brings the drain closer and closer to the source. The electrostatic design of MOSFETs has gotten more challenging as device dimensions have been scaled down over the past five decades, but the principles have not changed. The nature of electron transport in transistors has, however, changed considerably as transistors have become smaller and smaller. A proper treatment of transport in nanoscale transistors is essential to understanding these devices and will be our focus beginning in Lecture 7.

Finally, we note that the energy band diagrams that we have sketched are similar to the energy band diagrams for a bipolar transistor [1, 2]. In fact, the two devices both operate by controlling current by manipulating the height of an energy barrier [4]. The source of the MOSFET is analogous to the emitter of the BJT, the channel to the base of the BJT, and the drain to the collector of a BJT. This close similarity will prove useful in understanding the operation of short channel MOSFETs. In Lecture 8, we will briefly discuss bipolar transistors.

Lecture 3 Exercise: E-band diagrams for p-MOSFETs

Our discussion of MOSFETs in terms of energy band diagrams, has been for n-channel MOSFETs. In this exercise, we'll examine the corresponding energy band diagrams for p-channel MOSFETs. Note that energy band diagrams are plots of *electron* energy vs. position. Electron energy increases upwards in these plots. We think about p-channel MOSFETs in terms of holes. The hole energy increases *downward* in these plots. The starting point for these plots is the equilibrium energy band diagram for a p-channel MOSFET shown below.

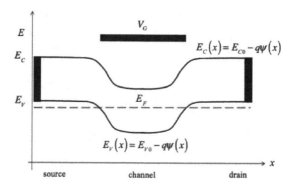

Equilibrium energy band diagram for a p-channel MOSFET. (Analogous to Fig. 3.3 for n-channel MOSFET.)

1) Sketch the energy band diagram for a PMOS transistor in the linear region.

Just as we only plotted $E_c(x)$ for NMOS transistors, we will only plot $E_v(x)$ for a PMOS transistor. The transistor is on, so $V_G < V_{TP} < 0$, and the gate voltage raises the electron energy in the channel (i.e. lowers the hole energy). The drain voltage is negative, so it raises F_p and E_v in the drain. The electric field in the channel is positive, which pushes holes from the source to the drain. The energy band diagram for the linear region PMOS transistor is shown below.

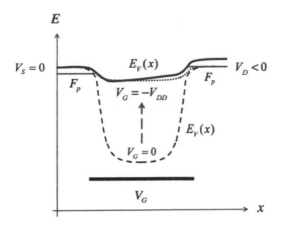

Energy band diagram for a p-channel MOSFET biased in the linear region. The dashed line is $E_v(x)$ in equilibrium. (Analogous to Fig. 3.4b for an n-channel MOSFET.)

2) Sketch the energy band diagram for a PMOS transistor in the on-state.

In this case, we are looking for the p-channel energy band diagram analogous to the on-state energy band diagram for the n-channel MOSFET, Fig. 3.5a. The result is shown below.

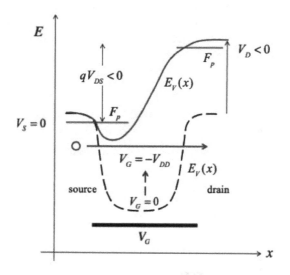

Energy band diagram for a p-channel MOSFET biased in the on-state. The dashed line is $E_v(x)$ in equilibrium. (Analogous to Fig. 3.5a for an n-channel MOSFET.)

3) Sketch the energy band diagram for a PMOS transistor in off-state.

We are looking here for the p-channel energy band diagram analogous to the off-state energy band diagram for the n-channel MOSFET, Fig. 3.5b. The result is shown below.

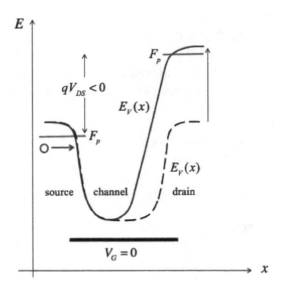

Energy band diagram for a p-channel MOSFET biased in the off-state. The dashed line is $E_v(x)$ in equilibrium. (Analogous to Fig. 3.5b for an n-channel MOSFET.)

3.11   References

Most of the important kinds of transistors are discussed in these texts:

[1] Y. Taur and T. Ning, *Fundamentals of Modern VLSI Devices*, 2nd Ed., Oxford Univ. Press, New York, 2013.

[2] Robert F. Pierret *Semiconductor Device Fundamentals*, 2nd Ed., Addison-Wesley Publishing Co., Reading, MA., 1996.

For a review of semiconductor fundamentals including the semiconductor equations and how current is related to gradients in the quasi-Fermi level, see:

[3] Robert F. Pierret *Advanced Semiconductor Fundamentals*, 2nd Ed., Pearson Education, Inc., 2003.

Johnson describes the close relation of bipolar and field-effect transistors.

[4] E.O. Johnson, "The IGFET: A Bipolar Transistor in Disguise," *RCA Review*, **34**, pp. 80-94, 1973.

Lecture 4

IV Theory: Traditional Approach

4.1   Introduction

MOSFET IV theory was developed in the 1960s to describe the first MOS-FETs and extended in the 1980s as channel lengths decreased below one micrometer to include velocity saturation. In this lecture we'll briefly review the traditional theory of the MOSFET as it is presented in textbooks (e.g. [1–3]). Only essential concepts will be discussed. For example, we will focus on the linear and saturation regions and not the entire IV characteristic, and only the above threshold IV characteristics will be discussed, not the important subthreshold characteristics. Those interested in a comprehensive treatment of traditional MOSFET theory should consult standard texts such as [4, 5]. In this lecture, we'll also introduce the *virtual source model*; in later lectures we'll show how it can be extended to describe the IV characteristics of modern, nanoscale transistors.

4.2   Current, charge, and velocity

Figure 4.1 shows the cross section of a planar MOSFET for which the drain to source current can be written as

$$I_D = W|Q_n(x)|\langle v_x(x)\rangle \qquad (4.1)$$

where W is the width of the transistor in the z-direction, Q_n is the mobile sheet charge in the $x-y$ plane (C/m^2), and $\langle v_x \rangle$ is the average x-directed velocity at which the charge flows. We assume that the device is uniform in the z-direction (out of the page) and that electrons flow in the x-direction from the source to the drain. In general, the mobile electron charge and velocity vary along the channel, but the current is constant if there is no recombination or generation. Accordingly, we can evaluate the current at the point along the channel where it is the most convenient to do so.

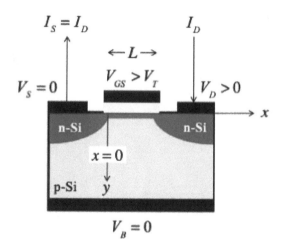

Fig. 4.1   Sketch of a planar, n-channel, enhancement mode MOSFET. The y-direction is normal to the channel, and the z-axis is out of the page. The beginning of the channel is located at $x = 0$. A mobile charge is present in the channel because $V_{GS} > V_T$; it is uniform between $x = 0$ and $x = L$ as shown here, when $V_S = V_D = 0$ and nearly uniform when V_{DS} is small.

Consider the MOSFET of Fig. 4.1 with $V_S = V_D = 0$, but with $V_{GS} > V_T$. The MOSFET is in equilibrium, so no current flows. In this case, the mobile charge is independent of x. As we will discuss in Lecture 5, there is very little charge in the channel when the gate voltage is less than the threshold voltage, V_T. For $V_{GS} > V_T$, the charge is negative and proportional to $V_{GS} - V_T$,

$$Q_n(V_{GS}) \approx -C_{ox}(V_{GS} - V_T) \quad \text{C/m}^2, \tag{4.2}$$

where C_{ox} is the gate *oxide capacitance* per unit area,

$$C_{ox} = \frac{\kappa_{ox}\epsilon_0}{t_{ox}} \quad \text{F/m}^2, \tag{4.3}$$

with the numerator being the dielectric constant of the oxide and the denominator the thickness of the oxide. (As we'll discuss in Lecture 5, the gate capacitance is actually somewhat less than C_{ox} when the oxide is thin.) Equation (4.2) reminds us of a simple, linear capacitor where $Q = CV$, except that it takes a finite voltage, V_T, to produce a significant mobile charge. To keep things simple in this lecture, we will assume that for $V_{GS} < V_T$, the charge is negligibly small.

When $V_D > V_S$, the mobile charge density varies with position along the channel, and so does the average velocity of electrons. In an electrostatically well-designed transistor, Q_n at the beginning of the channel (the top of the source to channel barrier, which we will refer to as the *virtual source*) is given by Eq. (4.2). Accordingly, we will evaluate I_D at $x = 0$, where we know the charge, and we only need to deduce the average velocity, $\langle v_x(x = 0)\rangle$.

4.3 Linear region

In the small V_{DS}, or linear region of the output characteristics (Fig. 1.6), a MOSFET acts as a voltage controlled resistor. Above threshold, the electric field in the channel is constant, and we can write the average velocity as

$$\langle v_x \rangle = -\mu_n \mathcal{E}_x = -\mu_n V_{DS}/L. \tag{4.4}$$

Using Eqs. (4.2) and (4.4) in (4.1), we find

$$\boxed{I_{DLIN} = \frac{W}{L}\mu_n C_{ox}(V_{GS} - V_T)V_{DS},} \tag{4.5}$$

which is the classic expression for the small V_{DS} drain current of a MOSFET. Note that we have labeled the mobility as μ_n, but in traditional MOS theory, this mobility is called the *effective mobility*, μ_{eff}. The effective mobility is the depth-averaged mobility in the inversion layer (see [1], p. 618). It is smaller than the electron mobility in the bulk, because *surface roughness scattering* at the oxide-silicon interface lowers the mobility.

4.4   Saturated region: Velocity saturation

In the large V_{DS}, or saturated region of the output characteristics, a MOS-FET acts as a voltage-controlled current source (Fig. 1.6). How do we compute I_{DSAT}? According to Eq. (4.4), the average electron velocity is proportional to the electric field, but this is only true when the electric field is small. Figure 4.2 is a sketch of the velocity vs. electric field for electrons in bulk Si. When $\mathcal{E} \ll \approx 10$ kV/cm, Eq. (4.4) holds, but when $\mathcal{E} \gg \approx 10$ kV/cm, the velocity saturates because the electrons gain kinetic energy from the strong electric field and move well above the bottom of the conduction band where the density-of-states is high and there are many more opportunities to scatter. As a result of strong carrier scattering, the electron velocity in bulk Si is limited to about 10^7 cm/s.

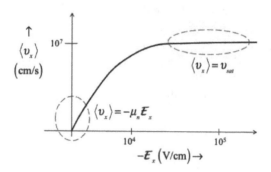

Fig. 4.2   Electron velocity saturation in bulk silicon. For small electric fields, $\langle v_x \rangle \propto \mathcal{E}_x$, but for large electric fields, $\langle v_x \rangle = v_{sat} = 10^7$ cm/s.

For a relatively small drain to source voltage of about 1 V, the electric field in the channel of a modern short channel ($L \approx 10$ nm) MOSFET is very high — well above the ≈ 10 kV/cm needed to saturate the velocity in bulk Si (Fig. 4.2). If the electric field is large across the entire channel for $V_{DS} > V_{DSAT}$, then the velocity is constant across the channel with a value of $v_{sat} = 10^7$ cm/s, and we can write the average velocity as

$$\langle v_x (x) \rangle = v_{sat} \approx 10^7 \quad \text{cm/s} \,. \tag{4.6}$$

Using Eqs. (4.2) and (4.6) in (4.1), we find

$$\boxed{I_{DSAT} = W C_{ox} v_{sat} \left(V_{GS} - V_T \right) \,,} \tag{4.7}$$

which is the *velocity saturated* drain current of a MOSFET. In practice, the current does not completely saturate, but increases slowly with drain voltage. For well-designed short channel Si MOSFETs, the output conductance is due to DIBL, i.e. the reduction of V_T with increasing drain voltage.

4.5   Saturated region: Classical pinch-off

For a long channel MOSFET under high drain bias, the electric field is moderate, and the velocity is not expected to saturate, but the drain current still saturates, so it must be for a different reason. This was the situation in early MOSFETs for which the channel length was about 10 micrometers (10,000 nanometers), and the explanation for drain current saturation was *pinch-off* near the drain, which is illustrated in Fig. 4.3.

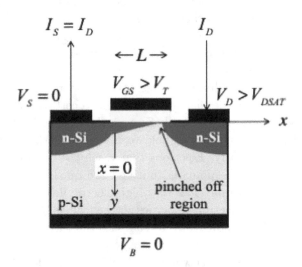

Fig. 4.3   Sketch of a long channel MOSFET showing the pinched-off region. Note that the thickness of the channel in this figure is used to illustrate the magnitude of the charge density (more charge near the source end of the channel than near the drain end). The channel is physically thin in the y-direction near the source end, where the gate to channel potential difference is large and physically thicker near the drain end, where the gate to channel potential difference is smaller. The length of the part of the channel where Q_n is substantial is $L' < L$.

Under high drain bias, the potential in the channel varies significantly from V_S at the source to V_D at the drain end (e.g. [7], pp. 51, 52). Since it is the difference between the gate voltage and the Si channel potential that matters, Eq. (4.2) must be extended as

$$Q_n(V_{GS}, x) = -C_{ox}\big(V_{GS} - V_T - \psi_s(x)\big),\qquad(4.8)$$

where $\psi_s(x)$ is the potential along the channel in the semiconductor. According to Eq. (4.8), when $\psi_s(L) = V_D = V_{GS} - V_T$, at the drain end, we find $Q_n(V_{GS}, L) = 0$. We say that the channel is *pinched off* at the drain. Of course, if $Q_n = 0$, then Eq. (4.1) states that $I_D = 0$, but a large drain current is observed to flow. This occurs because in the pinch-off region, carriers move very fast in the high electric field, so only a very small but non-zero Q_n is sufficient to carry the current. Over most of the channel, the electric field is moderate, and the mobile charge density is high, but the current saturates for $V_{DS} > (V_{GS} - V_T)$ because the additional voltage is dropped across the small, pinched-off part of the channel. The voltage drop across the conductive part of the channel remains at about $V_{GS} - V_T$. When carriers enter the pinched-off region near the drain, a large electric field quickly sweeps them across the pinched-off region and into the drain.

Some insight into pinch-off is provided by the "toy" numerical simulations shown in Fig. 4.4, which shows $E_c(x)$ under a high gate voltage for a series of increasing drain voltages. At first, increases in V_{DS} increase the electric field across the channel, but beyond a certain voltage (V_{DSAT}), increases in V_{DS} have only a small effect on the electric field to the left of the pinch-off point. The electric field to the right of the pinch-off point, however, increases greatly. Electrons that reach the pinch-off point don't stop; they fall down the potential hill and are quickly swept out of the channel and into the drain. The very high electric field in the pinched off region produces high electron velocities so that while $Q_n \approx 0$, the current is maintained. We are now ready to compute the saturated drain current due to pinch-off.

In the part of the channel where the mobile charge density is large, we can write the average velocity as

$$\langle v_x(x)\rangle = -\mu_n \mathcal{E}_x(x).\qquad(4.9)$$

The potential at the beginning of the channel is $\psi_s(0) = V_S = 0$, and the potential at the end of the channel where it is pinched off is $\psi_s(L') = V_{GS} - V_T$. The average electric field in the channel is the voltage across the non-pinched-off part of the channel, $V_{GS} - V_T$, divided by L', where

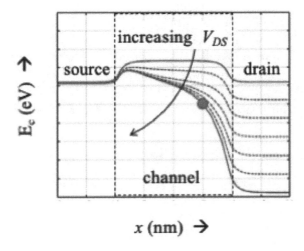

Fig. 4.4 Numerical simulations illustrating pinch-off in a MOSFET. The filled circle indicates the pinch-off point. For $V_{DS} > V_{DSAT}$, increases in drain voltage increase the electric field in the pinched-off portion of the channel but have little effect on the electric field in the part of the channel with a significant mobile charge density.

$L' < L$ is the length of the part of the channel that is not pinched off. The electric field at the beginning of the channel is

$$\mathcal{E}_x(0) = \frac{V_{GS} - V_T}{2L'}, \tag{4.10}$$

where the factor of two comes from a proper treatment of the nonlinear electric field in the channel (e.g. see [7], pp. 51, 52). Using Eq. (4.10) in (4.9), we find

$$\langle v_x(0) \rangle = -\mu_n \mathcal{E}_x(0) = -\mu_n \frac{V_{GS} - V_T}{2L'}. \tag{4.11}$$

Finally, using Eqs. (4.2) and (4.11) in (4.1), we find

$$\boxed{I_{DSAT} = \frac{W}{2L'}\mu_n C_{ox} \left(V_{GS} - V_T\right)^2,} \tag{4.12}$$

the so-called *square law IV characteristic* of a long channel MOSFET. In practice, the current does not completely saturate, but increases slowly with drain voltage as the pinched-off region slowly moves towards the source, which effectively decreases the length of the conductive part of the channel, L'.

Figure 4.5 compares the output characteristics of a long channel, square law MOSFET, with a short channel, velocity saturated MOSFET. The "signature" of a velocity saturated MOSFET is a saturation current that varies as $(V_{GS} - V_T)^1$. Early MOSFETs were square law devices for which the saturation current varied as $(V_{GS} - V_T)^2$. As channel lengths became shorter, the saturation current was found to vary as $(V_{GS} - V_T)^\alpha$, where $1 \leq \alpha \leq 2$, but today, most MOSFETs display the velocity saturated characteristic as illustrated by the typical IV measured characteristics in Fig. 3.6.

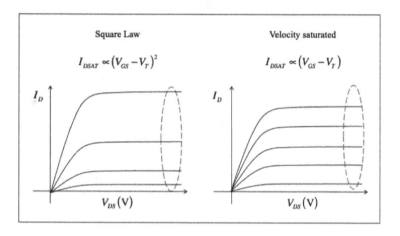

Fig. 4.5 Comparison of the output characteristics of square law and velocity saturated MOSFETs.

Although we have only discussed the linear and saturation region currents, it is possible to compute the square law I_D from small to large V_{DS} (see the exercise at the end of this lecture). It is also possible to compute I_D from small to large V_{DS} with the velocity saturation model [4, 5]. To keep the focus on essentials, we will not take the time to do this, but in Sec. 4.7, we will present a semi-empirical way to smoothly connect the linear and saturation region currents.

4.6 Series resistance

As illustrated in Fig. 4.6a, there is always some series resistance between the MOSFET contacts and the intrinsic device. Figure 4.6a shows how the voltages applied to the terminals of the device are related to the voltages on the intrinsic contacts. Here, V'_D, V'_S, and V'_G refer to the voltages on the terminals and V_D, V_S, and V_G refer to the voltages on the intrinsic terminals. (No resistance is shown in the gate lead, because we are considering DC operation now. Since the DC gate current is zero, a resistance in the gate has no effect. Gate resistance is, however, an important factor in the high frequency operation of transistors.)

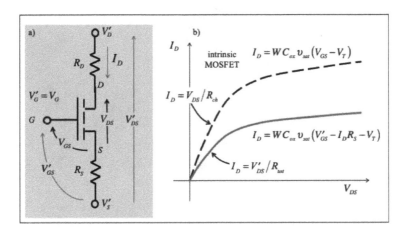

Fig. 4.6 Series resistances in a MOSFET. Left a): Relation of the voltages applied to the external contacts to the voltages on the internal contacts. Right b): The effect of series resistance on the *IV* characteristic. The dashed curve is an intrinsic MOSFET for which $R_S = R_D = 0$. As indicated by the solid line, series resistance increases the total resistance between the source and drain, R_{tot}, and lowers the saturation current.

From Fig. 4.6a, we relate the internal (unprimed) voltages to the external (primed) voltages by

$$V_G = V'_G$$
$$V_D = V'_D - I_D \left(V_G, V_S, V_D \right) R_D \, . \qquad (4.13)$$
$$V_S = V'_S + I_D \left(V_G, V_S, V_D \right) R_S$$

The *IV* characteristic of the intrinsic device, $I_D \left(V_G, V_S, V_D \right)$, are known, so Eqs. (4.13) are two equations in two unknowns — the internal voltages, V_D and V_S. Given applied voltages on the gate, source, and drain, V'_G, V'_S, V'_D, we can solve these equations for the internal voltages, V_S and V_D, and for the current, I_D.

Figure 4.6b illustrates the effect of series resistance on the *IV* characteristic. In the linear region, we can write the current of the intrinsic device as

$$I_{DLIN} = \frac{W}{L}\mu_n C_{ox}\left(V_{GS} - V_T\right)V_{DS} = V_{DS}/R_{ch}, \qquad (4.14)$$

but when source and drain series resistors are present, the linear region current becomes

$$I_{DLIN} = V_{DS}/R_{tot}, \qquad (4.15)$$

where

$$R_{tot} = R_{ch} + R_S + R_D. \qquad (4.16)$$

In the linear region, series resistance lowers the slope of the IV characteristic as shown in Fig. 4.6b.

Figure 4.6b also shows that series resistance decreases the saturation region current. In an ideal MOSFET with no output conductance, the drain series resistance has no effect in the saturation region when $V_D > V_{DSAT}$, but the source resistance reduces the intrinsic V_{GS}, so Eq. (4.7) becomes

$$I_{DSAT} = W C_{ox} \upsilon_{sat}\left(V'_{GS} - I_{DSAT}R_S - V_T\right). \qquad (4.17)$$

Series resistance lowers the internal gate to source voltage of a MOSFET, and therefore lowers the saturation current. The maximum voltage applied between the gate and source is the power supply voltage, V_{DD}. Series resistance will have a small effect on I_{DSAT} if $I_{DSAT}R_S \ll V_{DD}$. Modern Si MOSFETs deliver about 1 mA/μm of on-current at $V_{DD} = 1$ V. Accordingly, R_S must be much less than 1000 $\Omega - \mu$m; series resistances of about 100 $\Omega - \mu$m are needed. Although we will primarily be concerned with understanding the physics of the intrinsic MOSFET, we should be aware of the significance of series resistance when analyzing measured data.

4.7   Virtual Source model

In this lecture, we developed expressions for the linear and saturation region drain currents as:

$$I_{DLIN} = \frac{W}{L}\mu_n C_{ox}\left(V_{GS} - V_T\right)V_{DS} \qquad (4.5)$$

$$I_{DSAT} = W C_{ox} \upsilon_{sat}\left(V_{GS} - V_T\right) \qquad (4.7)$$

As shown in Fig. 4.7, these equations provide a rough description of I_D vs. V_{DS}, especially if we treat the output conductance by including DIBL according to

$$V_T(V_{DS}) = V_{T0} - \delta V_{DS}, \tag{4.18}$$

where V_{T0} is the threshold voltage for $V_{DS} = 0$, and δ is the DIBL parameter in units of V/V. If we now define the drain saturation voltage as the voltage where $I_{DLIN} = I_{DSAT}$, we find

$$V_{DSAT} = \frac{v_{sat}L}{\mu_n}. \tag{4.19}$$

For $V_{DS} \ll V_{DSAT}$, $I_D = I_{DLIN}$, and for $V_{DS} \gg V_{DSAT}$, $I_D = I_{DSAT}$.

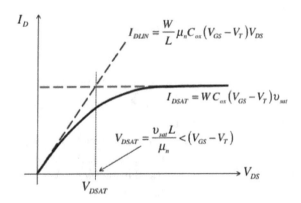

Fig. 4.7 Sketch of a common source output characteristic of an n-channel MOSFET at a fixed gate voltage (solid line). The dashed lines are the linear and saturation region currents as given by Eqs. (4.5) and (4.7).

Traditional MOSFET theory develops expressions for I_D vs. V_{DS} that smoothly transition from the linear to saturation regions as V_{DS} increases from zero to V_{DD} [1–5]. The goal in this section is to develop a simple, semi-empirical expression that describes the complete $I_D(V_{DS})$ characteristic from the linear to saturated region. The approach is similar to the so-called virtual source (VS) MOSFET model that has been developed and successfully used to describe a wide variety of nanoscale MOSFETs [6].

Equation (4.5) for the linear current can be written as

$$I_{DLIN}/W = |Q_n(V_{GS})| \times |\langle v_x(V_{DS})\rangle|$$
$$Q_n(V_{GS}) = -C_{ox}(V_{GS} - V_T) \tag{4.20}$$
$$\langle v_x(V_{DS})\rangle = -\mu_n V_{DS}/L.$$

Similarly, Eq. (4.7) for the saturation current can be written as

$$I_{DSAT}/W = |Q_n(V_{GS})| \times |\langle v_x(V_{DS})\rangle|$$
$$Q_n(V_{GS}) = -C_{ox}(V_{GS} - V_T) \tag{4.21}$$
$$\langle v_x(V_{DS})\rangle = -v_{sat}.$$

If we can find a way for the average velocity to go smoothly from its value at low V_{DS} to v_{sat} at high V_{DS}, then we will have a model that covers the complete range of drain voltages.

The VS model takes an empirical approach and writes the average velocity at the beginning of the channel as [6]

$$\langle v_x(V_{DS})\rangle = F_{SAT}(V_{DS})v_{sat}$$
$$F_{SAT}(V_{DS}) = \frac{V_{DS}/V_{DSAT}}{\left[1 + (V_{DS}/V_{DSAT})^\beta\right]^{1/\beta}}, \tag{4.22}$$

where V_{DSAT} is given by Eq. (4.19), and β is an empirical parameter chosen to fit the measured IV characteristic.

The form of the drain current saturation function, F_{SAT}, is motivated by the observation that the lower of the two velocities in Eqs. (4.20) and (4.21) should be the one that limits the current. We might, therefore, expect

$$\frac{1}{|\langle v_x(V_{DS})\rangle|} = \frac{1}{(\mu_n V_{DS}/L)} + \frac{1}{v_{sat}}, \tag{4.23}$$

which can be re-written as

$$|\langle v_x(V_{DS})\rangle| = \frac{V_{DS}/V_{DSAT}}{[1 + (V_{DS}/V_{DSAT})]}v_{sat}. \tag{4.24}$$

Equation (4.24) is similar to Eq. (4.22), except for the empirical parameter, β, which is adjusted to better fit data. Typical values of β for n- and p-channel Si MOSFETs are between 1.4 and 1.8 [6]. Equations (4.1), (4.2), and (4.22) give us a description of the above-threshold MOSFET for any drain voltage from the linear to the saturated regions.

Our simple VS model for the above threshold MOSFET is:

$$I_D/W = |Q_n(0)| \times |\langle v_x(0)\rangle|$$

$$Q_n(V_{GS}) = 0 \quad V_{GS} \leq V_T$$
$$Q_n(V_{GS}) = -C_{ox}(V_{GS} - V_T) \quad V_{GS} > V_T$$
$$V_T = V_{T0} - \delta V_{DS}$$

$$|\langle v_x(V_{DS})\rangle| = F_{SAT}(V_{DS})v_{sat}$$
$$F_{SAT}(V_{DS}) = \frac{V_{DS}/V_{DSAT}}{\left[1 + (V_{DS}/V_{DSAT})^\beta\right]^{1/\beta}}$$

$$V_{DSAT} = \frac{v_{sat}L}{\mu_n}$$

$$\text{(4.25)}$$

With this simple model, we can compute surprisingly good MOSFET *IV* characteristics with only a few device-specific model parameters: C_{ox}, V_T, μ_n, v_{sat}, L, and β. As discussed in Sec. 4.6, series resistance is important and is easily included, which adds another device-specific model parameter. This version of the model does not describe the subthreshold characteristics, but as we will see in the next lecture, that is readily included too.

4.8   Device metrics

Equations (4.5) and (4.7) can be used to relate some of the device metrics discussed in Lecture 2 to the underlying physics. Using these equations, we find:

$$I_{ON} = WC_{ox}v_{sat}(V_{DD} - V_T) \qquad V_T = V_{T0} - \delta V_{DS}$$
$$g_m^{sat} = \left.\frac{\partial I_{DS}}{\partial V_{GS}}\right|_{V_{GS}=V_{DS}=V_{DD}} = WC_{ox}v_{sat}$$
$$r_o = \left(\left.\frac{\partial I_{DS}}{\partial V_{DS}}\right|_{V_{GS}=V_{DD},V_{DS}>V_{DSAT}}\right)^{-1} = \frac{1}{g_m^{sat}\delta}$$
$$|A_v| = g_m^{sat}r_o = \frac{1}{\delta}$$

$$\text{(4.26)}$$

Equations (4.26) show that a high on-current and transconductance requires high capacitance and a high carrier velocity. High output resistance and self-gain require low DIBL. For analog electronics, channel lengths that are longer than the minimum in a given technology are used because longer channel MOSFETs have lower DIBL and, therefore, a higher self-gain.

Over many technology generations, the gate oxide thickness was scaled down to increase the gate capacitance and, therefore, the transconductance and on-current. Eventually, SiO_2 was replaced by gate dielectrics with higher dielectric constants. Carrier velocity is closely related to the mobility (even in the saturation region, as we will learn in Lecture 7 when we dive deeper into carrier transport). For Si MOSFETs, strain is used to increase the mobility. Field-effect transistors with III-V semiconductor channels have higher carrier velocities than Si MOSFETs because of the small effective masses of electrons in III-V semiconductors.

4.9   Discussion

The continuous down-scaling of channel lengths from one technology generation to the next required a deeper and deeper understanding of the physics of electron transport in short channels. For example, *diffusive transport* ($\langle v_x \rangle = -\mu_n \mathcal{E}_x$) only holds when the channel length is many mean-free-paths long. (The mean-free-path is the average distance an electron travels before it scatters from a lattice vibration, impurity in the lattice, roughness at the Si-oxide interface, etc.) Figure 4.8 is an example of a rigorous simulation of electron transport across a short channel transistor.

Figure 4.8a shows $E_c(x)$ for the on-state of an $L = 30$ nm MOSFET. Each dot in the figure represents an electron that is tracked by the computer as it accelerates in the electric field, scatters, is accelerated again, scatters again, until it eventually leaves the device from the source or drain contact and another electron is injected. Figure 4.8b shows the average x-directed, velocity vs. position, $\langle v_x(x) \rangle$ for the electrons in Fig. 4.8a. The first thing we notice is that the velocity in the channel does not saturate — it exceeds $v_{sat} = 10^7$ cm/s across the entire channel. This effect is known as *velocity overshoot* and occurs because the strong scattering that limits the high-field electron velocity in bulk Si cannot happen in the short time that it takes for an electron to zip across the channel. Velocity overshoot became important when channel lengths reached a few tenths of a micron. Today, channel lengths are about 0.01 micrometers (10 nm).

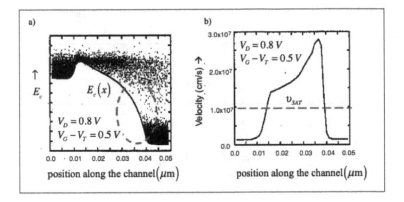

Fig. 4.8   Monte Carlo simulations of electron transport in an $L = 30$ nm MOSFET. Left a): $E_c(x)$ in the on-state showing the simulated electrons as points. Right b): The average electron velocity, $\langle v_x(x) \rangle$ along the channel. In the region near the drain indicated with the dashed oval, the electric field greatly exceeds the critical field for velocity saturation in bulk Si, but the velocity does not saturate. (©IEEE 1992. Reprinted, with permission, from: D. Frank, S. Laux, and M. Fischetti, Int. Electron Dev. Mtg., Dec., 1992.)

One might have expected that the velocity saturation model would be applicable only to MOSFETs with channel lengths longer than several hundred nanometers where velocity overshoot does not occur. Surprisingly, as shown in Fig. 4.9, we find that it accurately describes the *IV* characteristics of MOSFETs with channel lengths well below 100 nm. To achieve such fits, we must view the electron mobility and saturation velocity as empirical parameters that are adjusted to fit to measured data. We find that with relatively small adjustments in these parameters, excellent fits to most transistors can be achieved. (The adjustments are larger for III-V FETs.) The two adjusted parameters are called the *injection velocity*, v_{inj}, (the saturation velocity in the traditional model) and the *apparent mobility*, μ_{app}, (the real mobility in the traditional model). The fact that this simple model describes modern transistors so well, tells us that it captures something essential about the physics of MOSFETs. In Lecture 7, we will discuss a simple, clear physical interpretation of μ_{app} and v_{inj}.

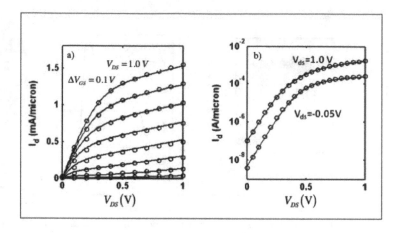

Fig. 4.9   Measured *IV* characteristics and fitted VS model calculations for 32 nm n-MOSFET technology. Left a): Common source output characteristic. Right b): Transfer characteristic. The VS model used for these fits is an extension of the model described by Eqs. (4.25) that uses an improved description of MOS electrostatics to treat the subthreshold as well as above threshold conduction. (©IEEE 2009. Reprinted, with permission, from: [6].)

4.10   Summary

In this lecture, we reviewed traditional MOSFET *IV* theory. In practice, there are several complications, but the essential features of the traditional approach are easy to grasp. We also recast the traditional MOSFET theory in the form of a simple virtual source model. The virtual source is located at the top of the source to channel barrier where $Q_n = -C_{ox}(V_{GS} - V_T)$. Application of the VS model to modern transistors shows that it describes them remarkably well. This is a consequence of the fact that it describes the essential features of a barrier-controlled transistor. The weakest part of the model is the description of electron transport, which is based on the use of a mobility and saturated velocity. Because of the simplified transport model, we need to regard the mobility and saturation velocity in the model as fitting parameters that are adjusted to fit experimental data.

According to Eq. (4.1), the drain current is proportional to the product of charge and velocity. The charge is controlled by MOS electrostatics (*i.e.* by using the gate voltage to modulate the energy barrier between the source and the channel). In the next lecture, we will discuss MOS electrostatics and learn how to describe subthreshold as well as above-threshold mobile charge. The result will be an improved VS model, but mobility and saturation velocity will still be viewed as fitting parameters. In Lectures 6 and 7, we'll discuss transport and learn how to formulate the VS model so that transport is described physically.

Lecture 4 Exercise: Square Law MOSFET IV Characteristic

In this lecture, we focused on deriving the linear and saturation region drain currents and then empirically connected them using the F_{SAT} function in the VS model. For a square law MOSFET, it is rather easy to derive expressions for I_D from $V_{DS} = 0$ to $V_{DS} > V_{DSAT}$ without the need for an empirical function.

Equations (4.5) and (4.12) describe the linear and saturation region currents as given by the square law theory of the MOSFET. In this exercise, we'll compute the complete IV characteristic from the linear region to the saturation region. We begin with Eq. (4.1) for the drain current and use Eq. (4.4) for the velocity to write

$$I_D = W|Q_n(x)|\langle v(x)\rangle = W|Q_n(x)|\mu_n \frac{d\psi_s}{dx}. \tag{4.27}$$

Next, we use Eq. (4.8) for the charge to write,

$$I_D = W\mu_n C_{ox}(V_{GS} - V_T - \psi_s(x))\frac{d\psi_s}{dx}, \tag{4.28}$$

then separate variables and integrate across the channel to find,

$$I_D \int_0^L dx = W\mu_n C_{ox} \int_{V_S}^{V_D} (V_{GS} - V_T - \psi_s)\,d\psi_s, \tag{4.29}$$

where we have assumed that I_D is constant (no recombination-generation in the channel) and that μ_n is constant as well. Integration gives us the IV characteristic of the MOSFET,

$$I_D = \frac{W}{L}\mu_n C_{ox}\left[(V_{GS} - V_T)V_{DS} - V_{DS}^2/2\right]. \tag{4.30}$$

Equation (4.30) gives the drain current for $V_{GS} > V_T$ and for $V_{DS} \leq (V_{GS} - V_T)$. For small V_{DS}, the result reduces to I_{DLIN} as given by Eq. (4.5). The region, $0 < V_{DS} < V_{GS} - V_T$ is known as the *triode region* of a MOSFET because of the similarity to the IV characteristics of a vacuum tube triode.

The charge in Eq. (4.8) goes to zero at $\psi_s(L) = V_{DS} = V_{GS} - V_T$, which defines the beginning of the pinch-off region,

$$V_{DSAT} = (V_{GS} - V_T). \tag{4.31}$$

The current beyond pinch-off for $V_{DS} > V_{DSAT}$ is found by evaluating Eq. (4.30) for $V_{DS} = V_{GS} - V_T$ and is

$$I_D = \frac{W}{2L'}\mu_n C_{ox}\left(V_{GS} - V_T\right)^2 , \qquad (4.32)$$

which increases slowly with V_{DS} because of channel length shortening due to pinch-off (i.e. $L' < L$), an effect that is known as *channel length modulation*.

Equations (4.30) and (4.32) give the square law *IV* characteristic of the MOSFET — not just the linear and saturated regions, but the entire *IV* characteristic.

4.11   References

The traditional theory of the MOSFET is described in the textbooks listed below.

[1] Robert F. Pierret *Semiconductor Device Fundamentals*, 2nd Ed., Addison-Wesley Publishing Co, 1996.

[2] Ben Streetman and Sanjay Banerjee, *Solid State Electronic Devices*, 6th Ed., Prentice Hall, 2005.

[3] J.A. del Alamo, *Integrated Microelectronic Devices*, Pearson, New York, 2018.

For comprehensive treatments of classical MOSFET theory, see:

[4] Y. Tsividis and C. McAndrew, *Operation and Modeling of the MOS Transistor*, 3rd Ed., Oxford Univ. Press, New York, 2011.

[5] Y. Taur and T. Ning, *Fundamentals of Modern VLSI Devices*, 2nd Ed., Oxford Univ. Press, New York, 2013.

The MIT Virtual Source Model is described in:

[6] A. Khakifirooz, O.M. Nayfeh, and D.A. Antoniadis, "A Simple Semiempirical Short-Channel MOSFET Current-Voltage Model Continuous Across All Regions of Operation and Employing Only Physical Parameters," *IEEE Trans. Electron. Dev.*, **56**, pp. 1674-1680, 2009.

For a more complete discussion of the approach to transistors used in these lectures, see:

[7] Mark Lundstrom, *Fundamentals of Nanotransistors*, World Scientific Publishing Co., Singapore, 2018.

Lecture 5

Mobile Charge

5.1 Introduction

The mobile charge that flows across the channel is controlled by the gate voltage, which lowers the barrier so that mobile charge carriers can enter the channel. Our goal in this lecture is to relate the mobile charge density in the channel, Q_n in C/m^2 for n-channel MOSFETs, to the gate voltage. Below threshold, $Q_n \propto \exp[q(V_{GS} - V_T)/mk_BT]$ and above threshold $Q_n \propto (V_{GS} - V_T)$. In this lecture, we'll show where these expressions come from.

In Lecture 4 we summarized the traditional theory of the MOSFET IV characteristics for a bulk (planar) MOSFET for which the channel is on the surface of a doped semiconductor that can be considered to be infinitely thick. For many years, all MOSFETs were of this kind, but today, the most advanced digital technologies make use of non-planar MOSFETs such as the FinFET (recall Fig. 3.11). In this lecture, we will discuss a model structure representative of non-planar MOSFETs. Those interested in a similar treatment for bulk MOSFETs should see Lecture 8 of [1].

Figure 5.1 is a double gate MOSFET that is similar to a tall FinFET in which the gate is wrapped around three sides of the channel or to a wide nanosheet MOSFET in which the gate is completely wrapped around the channel. The Si layer is physically thin in the y-direction (typically a few nanometers), and the mobile charge lies in the $x - z$ plane. The channel is undoped, or if lightly doped it is fully depleted. The energy band diagram along the channel, $E_c(x)$, looks just like the figures in Lecture 3. In this lecture, we'll draw energy band diagrams in the y-direction, normal to the channel, in order to examine how the gate voltage controls Q_n. We'll consider only the equilibrium problem ($V_{DS} = 0$), but in an electrostatically well-designed MOSFET, the results also apply under high V_{DS} at the virtual source, which is located at the top of the barrier. Finally, to keep the discussion focussed on essentials, we'll assume nondegenerate carrier statistics and gloss over some details about quantum confinement and bandstructure; interested readers should refer to Lecture 9 in [1].

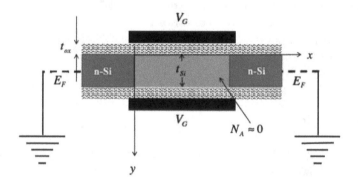

Fig. 5.1   A model double gate MOSFET structure used in this lecture to illustrate MOS electrostatics for non-planar MOSFETs. The z-axis is into the page. Since $V_S = 0, V_G = V_{GS}$.

5.2   The mobile charge

When electrons are confined in a thin semiconductor film, they behave as quantum mechanical particles in a box. Electrons are free to move in the $x - z$ plane, but confinement in the y-direction causes the conduction band to become a set of *subbands* with the bottom of each subband being a particle in a box energy. These *quasi-two-dimensional* electrons are described by a two-dimensional density-of-states [1, 2]

$$D_{2D} = \frac{m_n^*}{\pi \hbar^2} \quad J^{-1}m^{-2}, \tag{5.1}$$

where m_n^* is the *density-of-states effective mass*. Recall that the constant energy surfaces of the conduction band of Si are ellipsoids described by the longitudinal effective mass, m_l^*, and the transverse effective mass, m_t^* [2]. For channels in the (100) plane, the lowest energy subband (the only one that we will be concerned with), the confinement energy is determined by m_l^*, and $m_n^* = 2m_t^*$. The factor of two comes from the fact that two of the six conduction band ellipsoids contribute identically to the first confined subband [1]. For a discussion of the band structure of bulk Si, how it changes due to quantum confinement, and the 2D density-of-states, see [1, 2].

The number of states per m^2 in the energy range, dE, is $D_{2D}(E)dE$, and the probability that these states are occupied is given by the equilibrium Fermi function,

$$f_0(E) = \frac{1}{1 + e^{(E - E_F)/k_B T}}. \tag{5.2}$$

The sheet carrier concentration per m^2, n_s, is the electron density per m^3 integrated across the thickness of the semiconductor; it is also the integral of the occupied 2D states in the conduction band:

$$\begin{aligned} n_s &= \int_0^{t_{Si}} n(y)dy = \int_0^{\infty} D_{2D}(E)f_0(E)dE \\ &= N_{2D} \ln\left(1 + e^{(E_F - E_c')/k_B T}\right) \quad m^{-2}, \end{aligned} \tag{5.3}$$

where

$$N_{2D} \equiv \frac{m_n^* k_B T}{\pi \hbar^2} \quad m^{-2}, \tag{5.4}$$

is the 2D *effective density-of-states* for the conduction band, and

$$E_c' = E_c + \epsilon_1, \tag{5.5}$$

is the bottom of the conduction band plus the confinement energy for the first subband, ϵ_1. (To keep things simple, we assume that only one subband is occupied.) We have also extended the integral in Eq. (5.3) to infinity because the Fermi function will assure that the integrand approaches zero before the top of the conduction band is reached.

For a nondegenerate semiconductor, $E_F < E'_c$, and the exponential in Eq. (5.3) is small. Using $\log(1 + x) \approx x$ for small x, we see that for a nondegenerate semiconductor, Eq. (5.3) becomes

$$n_s = N_{2D} e^{(E_F - E'_c)/k_B T} . \tag{5.6}$$

The two gates shown in Fig. 5.1 are used to change the potential, ψ_s, in the semiconductor. A symmetrical, double gate structure is assumed with the same voltage applied to the top and bottom gates. The Fermi level is the same in the source, channel, and drain because $V_S = V_D = 0$ is assumed. We also assume that the Si layer is thin enough and the electron density small enough so that the bottom of the conduction band is nearly flat, which means that the electrostatic potential in the semiconductor, ψ_s, is not a function of y. With these assumptions, we can write

$$E'_c = E'_{co} - q\psi_s , \tag{5.7}$$

where E'_{co} is E'_c when $\psi_S = 0$. Using Eq. (5.7), Eq. (5.6) becomes.

$$\begin{aligned} n_s(\psi_s) &= N_{2D} e^{(E_F - E'_{co} + q\psi_s)/k_B T} \\ &= \left(N_{2D} e^{(E_F - E'_{co})/k_B T} \right) e^{q\psi_s/k_B T} , \\ &= n_{so} e^{q\psi_s/k_B T} \end{aligned} \tag{5.8}$$

where n_{so} is the sheet electron density when $\psi_s = 0$, which depends, of course, on the location of the Fermi level. An analogous treatment for mobile holes in p-channel MOSFETs can also be done The Fermi level is typically placed in the middle of the bandgap because a mid-gap Fermi level makes $V_{TN} = -V_{TP}$. In this case, $E_F = E_i$, and $n_{so} = n_{si}$, where n_{si} is the intrinsic, 2D sheet carrier concentration.

Our final expression for the mobile electron charge is

$$\boxed{Q_n(\psi_s) = -q n_{si}\, e^{q\psi_s/k_B T} \ \text{C/m}^2 .} \tag{5.9}$$

Equation (5.9) is valid below and above threshold. It does assume nondegenerate carrier statistics, which introduces errors above threshold, but the general concepts (our focus in these lectures) are most clearly illustrated by using the nondegenerate expressions. Those interested in a more comprehensive treatment that includes nondegenerate carrier statistics, quantum confinement, and bandstructure effects, should consult Lecture 9 in [1].

Now that we know how the mobile charge depends on the potential in the semiconductor, the next question is: "How does the mobile charge depend on the gate voltage?" Figure 5.2 illustrates qualitatively how the gate voltage affects the energy bands. For positive gate voltage, the potential in the semiconductor increases, E'_{co} moves down — closer to the Fermi level, and the electron concentration increases exponentially. For a negative gate voltage, the valence band moves up, and the hole concentration increases exponentially. In the next section, we discuss the quantitative relation between V_{GS} and ψ_s.

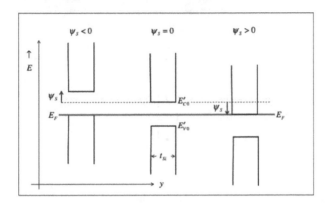

Fig. 5.2   Illustration of how a negative, zero, and positive electrostatic potential, ψ_s, affects the energy band diagram of the double gate MOSFET of Fig. 5.1.

5.3   Semiconductor potential vs. gate voltage

Now that we understand how the mobile charge is related to the potential in the semiconductor, we need to understand how ψ_s is related to the gate voltage. First, let's refresh our memory on Gauss's Law, which states that the electric flux (or displacement flux, D) passing through any closed surface (a *Gaussian surface*) is equal to the total charge enclosed by that surface. Mathematically, we write

$$\oint_S \vec{D} \cdot d\vec{S} = Q_{tot}, \qquad (5.10)$$

where $\vec{D}$ is related to the electric field by $\vec{D} = \epsilon \vec{\mathcal{E}}$ and $d\vec{S}$ is a vector representing a differential area of the surface with a direction in the outward normal to the surface. Consider the two cases shown in Fig. 5.3.

At the top, in Fig. 5.3a, is a semi-infinite p-type semiconductor that is neutral except for a sheet of negative charge, Q_n, in C/m², on the left surface. A cylindrical Gaussian surface is also shown. From the symmetry

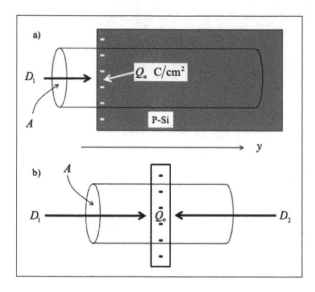

Fig. 5.3   Illustration of Gauss's Law. Top a): A semi-infinite layer of p-type Si that is neutral except for a sheet of negative charge at the left surface. Bottom b): A thin layer of undoped Si with a sheet charge of $Q_n < 0$ in the layer.

of the problem, we see that $\vec{D}$ points in the y-direction. The only contribution to the closed integral comes from the left surface of area, A, because the semiconductor is neutral, so $\vec{D} = 0$ on the other face and $\vec{D}$ and $d\vec{S}$ are perpendicular on the rest of the surface. We conclude that

$$-D_1 A = Q_{tot} = Q_n A$$

or $D_1 = -Q_n$. The displacement field normal to the surface is minus the surface charge density.

Now consider a thin Si film with a negative charge Q_n, distributed uniformly within it as shown in Fig. 5.3b. This is the case for our thin Si channel. In this case, we find for the y-directed displacement field

$$D_1 = -D_2 = -Q_n/2 \,.$$

The displacement field normal to the surface is minus one-half of the surface charge density.

With this review, we can now relate the potential in the semiconductor to the gate voltage. Figure 5.4 shows energy band diagrams in the y-direction for the case of $V_{GS} = 0$ (Fig. 5.4a) and $V_{GS} > 0$ (Fig. 5.4b).

Figure 5.4a assumes that the *workfunctions* of the metal gates are such that the Fermi level in the metal gates lines up with the intrinsic Fermi

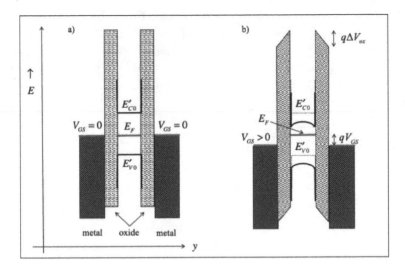

Fig. 5.4   Energy vs. position in the direction normal to the plane of the channel for the double gate MOSFET. Left a): Zero gate voltage. Right b): Positive gate voltage.

level in the semiconductor, so $n_s = p_s = n_{si}$ (the intrinsic, 2D sheet carrier concentration), and there is no charge in the semiconductor. As a result, there is no band banding at $V_{GS} = 0$ V. The reference for potential is arbitrary, so we define $\psi_s = 0$ under these conditions.

When a positive gate voltage is applied, the Fermi level in the two metal gates is pulled down by qV_{GS} as shown in Fig. 5.4b. The Fermi level in the semiconductor does not change because it is determined by the source and drain contacts, and $V_S = V_D = 0$; the semiconductor is still in equilibrium. The positive gate voltage produces a positive potential in the semiconductor, $\psi_s > 0$, which lowers the bands bringing E'_{co} closer to E_F, which increases $|Q_n|$. Because there is now negative charge in the semiconductor, there will be a $\vec{D}$-field on each side of the semiconductor, just as in Fig. 5.3b. As a result, there is an electric field in the oxide and as shown in Fig. 5.4b, a potential drop of ΔV_{ox} across the oxide. If there is no charge at the oxide-Si interface, then the $\vec{D}$-field must be continuous, so if there is an electric field in the oxide, there must also be an electric field in the Si — as shown by the band bending in Fig. 5.4b. This means that ψ_s in the semiconductor is a function of y. If the Si layer is thin, however, the potential change is small, so we will continue to treat ψ_s as constant across the the thickness of the Si layer.

We can now relate ψ_s to the gate voltage. From the symmetry of the double gate structure, we see that half of the negative charge in the semiconductor will image on a positive sheet charge on the left gate and the other half on the right gate. Because of this symmetry, we only need to relate the voltage on the left gate to the charge in the left half of the channel. Kirchoff's Voltage Law gives

$$V_{GS} = \Delta V_{ox} + \psi_s \,. \tag{5.11}$$

We find the electric field in the oxide from Gauss's Law,

$$D_{ox} = \kappa_{ox}\epsilon_0\mathcal{E}_{ox} = -\frac{Q_n(\psi_s)}{2} \,, \tag{5.12}$$

where κ_{ox} is the relative dielectric constant of the oxide, and ϵ_0 is the permittivity of free space. The potential drop across the oxide is

$$\Delta V_{ox} = \mathcal{E}_{ox}t_{ox} \,, \tag{5.13}$$

so from Eqs. (5.11)–(5.13), we find

$$\boxed{V_{GS} = -\frac{Q_n(\psi_s)}{C_{ox}} + \psi_s \,,} \tag{5.14}$$

where

$$C_{ox} = 2\left(\frac{\kappa_{ox}\epsilon_0}{t_{ox}}\right) \qquad \text{F/m}^2 \,, \tag{5.15}$$

is the total oxide capacitance per unit area of the two gates.

Equation (5.14) is an important result that relates ψ_s to V_{GS}. Using Eq. (5.9) in (5.14), we find

$$V_{GS} = \left[\frac{qn_{si}\,e^{q\psi_s/k_BT}}{C_{ox}}\right] + \psi_s \,. \tag{5.16}$$

If we could solve this equation for ψ_s in terms of V_{GS} and insert the result in Eq. (5.9), we would have the expression that we are looking for, $Q_n(V_{GS})$; unfortunately, it is not possible to analytically solve Eq. (5.16) for ψ_s. It is possible, however to plot ψ_s vs. V_{GS} by assuming a ψ_s and then computing the V_{GS} that produced it. As shown in Fig. 5.5, below threshold, $\psi_s = V_{GS}$, but above a certain ψ_s, the semiconductor potential varies much more slowly with V_{GS}.

' Equation (5.14) explains the shape of the $\psi_s(V_{GS})$ characteristic. Below threshold, Q_n is small, so the voltage drop across the oxide is negligible and Eq. (5.14) gives $\psi_s = V_{GS}$, but we see in Fig. 5.5 that above a certain

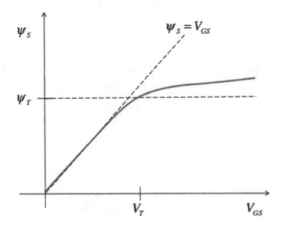

Fig. 5.5   Sketch of semiconductor potential, ψ_s, vs. gate voltage, V_{GS}. Also shown as dashed lines are $\psi_s = V_{GS}$ and $\psi_s = \psi_T$. The approximate gate threshold voltage, V_T, is also shown. The sketch assumes $m \approx 1$, as it is for a long channel, fully-depleted MOSFET or for a well-designed short channel MOSFET.

ψ_s labeled ψ_T, it becomes very difficult to increase ψ_s by increasing V_{GS}. Equation (5.14) explains why. When $\psi_s = \psi_T$, $|Q_n|$ has become substantial and the voltage drop across the oxide can no longer be ignored. A small increase in ψ_s produces an exponential increase in the voltage drop across the oxide, the term in square brackets in Eq. (5.16). If it takes a critical gate voltage, V_T, to reach $\psi_s = \psi_T$, then an increase in gate voltage above V_T mostly increases the voltage drop across the oxide and has little effect on ψ_s.

The potential in the semiconductor varies linearly with V_{GS} below threshold, how does it vary above threshold? Let's assume that $\psi_s \approx \psi_T$ above threshold, move it to the LHS in Eq. (5.16), replace it by V_T, the gate voltage at $\psi_s = \psi_T$, and then solve for ψ_s. We find

$$\psi_s = (k_B T/q) \ln\left[\left(\frac{C_{ox}}{q n_{si}}\right)(V_{GS} - V_T)\right], \qquad (5.17)$$

so ψ_s varies linearly with V_{GS} below threshold and as the logarithm of $(V_{GS} - V_T)$ above threshold. We now understand how Q_n is related to ψ_s according to Eq. (5.9) and how ψ_s is related to V_{GS} as given by Eqs. (5.14) and (5.16) and summarized in Fig. 5.5. In the next two sections, we'll examine $Q_n(V_{GS})$.

5.4   The mobile charge below threshold

We saw in the previous section that below threshold, $\psi_s = V_{GS}$, but as discussed in Sec. 3.7, Eq. (3.12), $\psi_s = V_{GS}/m$ where $m \geq 1$ because the gate does not have complete control over the potential at the top of the barrier — the drain has some influence too. Nonplanar MOSFETs were adopted to increase the electrostatic control by the gate so that m is as close to one as possible. It is now easy to convert Eq. (5.9) to an expression for $Q_n(V_{GS})$ below threshold,

$$Q_n(V_{GS}) = -qn_{si}\,e^{qV_{GS}/k_B T} . \tag{5.18}$$

This equation applies for $V_{GS} < V_T$, where V_T is the gate voltage that produced $\psi_s = \psi_T$, but what is the value of ψ_T? We might argue that when the conduction band has been pulled down so that $E_F = E_c'$, then the electron concentration will become significant. Equation (5.6) shows that when this condition holds and when non-degenerate statistics are used, then $n_s = N_{2D}$. Accordingly, we can find the semiconductor potential at threshold, ψ_T, from

$$n_s(\psi_T) = n_{si}\,e^{q\psi_T/k_B T} = N_{2D} , \tag{5.19}$$

which gives the potential at threshold as

$$\psi_T = \frac{k_B T}{q}\ln\left(\frac{N_{2D}}{n_{si}}\right) . \tag{5.20}$$

At threshold, the charge in the semiconductor is still small so the voltage drop across the oxide can be ignored to write

$$V_T = \psi_T = \frac{k_B T}{q}\ln\left(\frac{N_{2D}}{n_{si}}\right) . \tag{5.21}$$

Equation (5.21) can be solved for n_{si}, which can be used in Eq. (5.18), to write the mobile charge below threshold as

$$\boxed{Q_n(V_{GS}) = -qN_{2D}\,e^{q(V_{GS}-V_T)/mk_B T} .} \tag{5.22}$$

Equation (5.52) is a key result, which shows that the subthreshold mobile charge increases exponentially with gate voltage. Note that in Eq. (5.22), we have inserted the parameter, $m > 1$, discussed in Sec. 3.7. This parameter accounts for the fact that with 2D electrostatics, the gate voltage only has partial control over the potential at the top of the barrier.

5.5 The mobile charge above threshold

We began this lecture by pointing out that below threshold, $Q_n \propto \exp[q(V_{GS} - V_T)/mk_BT]$ and above threshold $Q_n \propto (V_{GS} - V_T)$. We now understand $Q_n(V_{GS})$ when $V_{GS} < V_T$; in this section, we'll discuss $Q_n(V_{GS})$ for $V_{GS} > V_T$.

Let's begin by re-writing Eq. (5.14) as

$$Q_n = -C_{ox}(V_{GS} - \psi_s) . \tag{5.23}$$

As shown in Fig. 5.5, $\psi_s \approx \psi_T$ above threshold, so $Q_n = -C_{ox}(V_{GS} - \psi_T)$ above threshold. At the onset of threshold, $Q_n(\psi_s = \psi_T) \approx 0$, so according to Eq. (5.14), $V_{GS} = V_T = \psi_T$, and we conclude that above threshold

$$Q_n \approx -C_{ox}(V_{GS} - V_T) , \tag{5.24}$$

which is the expression we used in Lecture 4 as we developed the traditional theory of the MOSFET. For many years, Eq. (5.24) was a good approximation, but as discussed next, it is no longer adequate because it does not give the correct gate capacitance.

The gate capacitance is $dQ_g/dV_{GS} = d(-Q_n)/dV_{GS}$. (Recall that $C = Q/V$ for a linear capacitor and $C = dQ/dV$ for a nonlinear capacitor.) From Eq. (5.24), we find the gate capacitance as

$$C_g \equiv \frac{d(-Q_n)}{dV_{GS}} \approx C_{ox} . \tag{5.25}$$

We find the actual gate capacitance from Eq. (5.23) as

$$\begin{aligned}
C_g = \frac{d(-Q_n)}{dV_{GS}} &= C_{ox}\left(1 - \frac{d\psi_s}{dV_{GS}}\right) \\
&= C_{ox}\left(1 - \frac{d\psi_s}{dQ_n}\frac{dQ_n}{dV_{GS}}\right) , \\
&= C_{ox}\left(1 + \frac{C_s}{C_g}\right)
\end{aligned} \tag{5.26}$$

where we have defined the *semiconductor capacitance* as

$$C_s \equiv \frac{d(-Q_n)}{d\psi_s} . \tag{5.27}$$

Equation (5.26) can be solved for the gate capacitance to find

$$C_g = \frac{C_{ox}C_s}{C_{ox} + C_s} < C_{ox} , \tag{5.28}$$

which we recognize as the series combination of two capacitors, the oxide capacitance and the semiconductor capacitance.

The semiconductor capacitance is very large because Q_n depends exponentially on ψ_s according to Eq. (5.9). A large capacitor in series with a much smaller capacitor has little effect, but after decades of scaling down the oxide thickness and finally replacing SiO_2 with oxides with higher dielectric constants, C_{ox} has become so large that C_s can no longer be ignored.

To develop an approximate expression for Q_n above threshold, we begin with Eq. (5.23), approximate ψ_s by V_T, which is independent of V_{GS}, and then replace C_{ox} out front with C_g as given by Eq. (5.28) to find

$$\boxed{Q_n = -C_g \left(V_{GS} - V_T \right).} \qquad (5.29)$$

In contrast to the approximation, Eq. (5.24), this approximation give the correct gate capacitance. Equation (5.29) is a key result, which shows that the above threshold mobile charge varies linearly with gate voltage.

5.6 Discussion

We have developed expressions for $Q_n(V_{GS})$ below and above threshold, but to model the complete behavior of a transistor, we need a single expression that covers the full range of gate voltages. As mentioned earlier, if we could solve Eq. (5.16) for ψ_s in terms of V_{GS} and insert the result in Eq. (5.9), we would have a full range expression for $Q_n(V_{GS})$; unfortunately, it is not possible to analytically solve Eq. (5.16) for ψ_s.

Note that the function, $\ln(1+e^x)$, which appears in the expression for n_s, Eq. (5.3), behaves properly in the two limits. For small e^x, $\ln(1 + e^x) \approx e^x$ and for large e^x, $\ln(1 + e^x) \approx x$. An empirical expression for Q_n using this function is [3]

$$Q_n(V_{GS}) = -mC_g \frac{k_b T}{q} \ln\left(1 + e^{q(V_{GS}-V_T)/mk_B T}\right). \qquad (5.30)$$

For $V_{GS} \gg V_T$, this expression reduces to

$$Q_n(V_{GS}) = -C_g(V_{GS} - V_T),$$

which is the correct result. For $V_{GS} \ll V_T$, we can use $\ln(1+x) \approx x$ to find

$$Q_n(V_{GS}) = -mC_g \frac{k_b T}{q} e^{q(V_{GS}-V_T)/mk_B T}.$$

Comparison with Eq. (5.22) shows that this is not the correct result — the pre-exponential factor is different. In practice, this is not so important because the subthreshold Q_n is dominated by the exponential factor, so the empirical expression works well. If we use Eq. (5.30) for $Q_n(V_{GS})$ in Eqs. (4.25), which describe the VS model, we now have a model that describes MOSFETs from the linear to saturation region and from subthreshold to above threshold. The treatment here is a simple version of The MIT Virtual Source model, which uses an extended version of Eq. (5.30) [4].

Finally, we should comment on the gate capacitance in the Q_n expression. The gate capacitance is the series combination of C_{ox} and the semiconductor capacitance, C_s. Evaluating C_s by differentiating Eq. (5.9), we see that it is proportional to Q_n, which depends on the gate voltage, so C_g depends on V_{GS}. We typically evaluate C_g under on-state conditions and assume that it is constant above threshold.

5.7 Summary

In this lecture, we have discussed how Q_n varies with surface potential and with gate voltage, considering both the subthreshold and above threshold regions. The important results are summarized in the following four equations.

$$Q_n(\psi_s) = -qn_{si}\, e^{q\psi_s/k_B T} \;\; \text{C/m}^2 \tag{5.9}$$

$$V_{GS} = -\frac{Q_n(\psi_s)}{C_{ox}} + \psi_s \tag{5.14}$$

$$Q_n(V_{GS}) = -qN_{2D}\, e^{q(V_{GS}-V_T)/mk_B T} \quad (V_{GS} < V_T) \tag{5.22}$$

$$Q_n = -C_g\,(V_{GS} - V_T) \quad\quad\quad\quad (V_{GS} > V_T) \tag{5.29}$$

According to Eq. (5.9), the mobile charge varies exponentially with the potential in the semiconductor. Equation (5.9) applies both below and above threshold. Equation (5.14) relates the semiconductor potential to the gate voltage. Below threshold, the mobile charge varies exponentially with gate voltage as described by Eq. (5.22), and above threshold, the mobile charge varies linearly with gate voltage as described by Eq. (5.29).

In the next two lectures, we'll develop a simple theory for the *IV* characteristics of nanoscale MOSFETs by evaluating Eq. (4.1),

$$I_D = W|Q_n\,(x = 0)|\,\langle v_x\,(x = 0)\rangle\,,$$

at the virtual source. In an electrostatically well-designed MOSFET, the expressions developed in this lecture for Q_n apply (with a small correction for DIBL) to the virtual source.

Lecture 5 Exercise: Gate capacitance above threshold

For many years, the gate oxide of MOSFETs was SiO_2 with a relative dielectric constant of $\kappa_{ox} = 4.0$. To increase the gate capacitance, modern MOSFETs make use of *high-k gate dielectrics* with a dielectric constant greater than SiO_2. The gate oxide is succinctly described in terms of its *equivalent oxide thickness, EOT*, which is the thickness of SiO_2 that would produce the same oxide capacitance. For this exercise, assume a MOSFET with a gate insulator having an *EOT* of 0.7 nm. Note that a 0.7 nm SiO_2 oxide would be too thin. Electrons would quantum mechanically tunnel through it, and the gate leakage current would be excessive. It was for this reason that high-k gate dielectrics were developed. Because they have a higher dielectric constant, thicker layers can be used while still achieving high gate capacitance.

This exercise is about the gate capacitance, C_g, which determines the above-threshold mobile charge according to Eq. (5.29)

$$Q_n = -C_g(V_{GS} - V_T).$$

According to Eq. (5.28),

$$C_g = \frac{C_{ox}C_s}{C_{ox} + C_s} < C_{ox},$$

so we need to compute the oxide capacitance and the semiconductor capacitance. The oxide capacitance is

$$C_{ox} = \frac{\kappa_{ox}\epsilon_0}{EOT} = 3.9 \times 10^{-6} \text{ F/cm}^2 = 3.9 \text{ } \mu\text{F/cm}^2.$$

The semiconductor capacitance is given by Eq. (5.27) as

$$C_s \equiv \frac{d(-Q_n)}{d\psi_s} = \frac{d(qn_S)}{d\psi_s},$$

where the sheet electron concentration is given by Eq. (5.3) as

$$n_s = N_{2D} \ln\left(1 + e^{\eta_c}\right) \text{ cm}^{-2},$$

with

$$N_{2D} \equiv \frac{m_n^* k_B T}{\pi \hbar^2} = 4.1 \times 10^{12} \text{ cm}^{-2},$$

and

$$\eta_c = (E_F - E_c')/k_B T = (E_F - E_{co} + q\psi_s)/k_B T \,.$$

To evaluate the effective density-of-states, N_{2D}, a density-of-states effective mass of $m_n^* = 2 \times 0.19 m_0$ was assumed. Here, $0.19 m_0$ is the transverse effective mass of electrons in the conduction band, and the factor of two is for the two valleys. This effective mass is appropriate for the lowest subband of (100) Si. See Sec. 9.2 in [1] for more discussion.

Calculations are always easier in the non-degenerate limit where Eq. (5.3) reduces to Eq. (5.6)

$$n_s = N_{2D} e^{\eta_c} \text{ cm}^{-2} \,,$$

from which we find the semiconductor capacitance as

$$C_s = \frac{q^2 n_s}{k_b T} = 62 \,\mu\text{F}/\text{cm}^2 \,.$$

A value of $n_s = 10^{13}$ cm^{-2}, which is appropriate for the on-state, has been assumed. Knowing the semiconductor capacitance and the oxide capacitance, we find the gate capacitance as

$$C_g = \frac{C_{\text{ox}} C_s}{C_{\text{ox}} + C_s} = 3.7 \,\mu\text{F}/\text{cm}^2 \,.$$

We conclude that the gate capacitance is roughly 5 percent less than the oxide capacitance for this specific oxide capacitance.

While the calculations are easier when non-degenerate carrier statistics are assumed, this is a case where we should be more careful because the non-degenerate assumption is not valid above threshold where we are computing C_g. Using Eq. (5.3) for n_s, we find

$$C_s = \frac{q^2 N_{2d}}{k_b T} \left(\frac{1}{1 + e^{-\eta_c}} \right) \,.$$

To evaluate the semiconductor capacitance, we need η_c. Since we know N_{2D} and n_s in the on-state, we can solve Eq. (5.3) and find

$$\eta_c = \ln\left(e^{n_s/N_{2D}} - 1\right) = 2.35 \,,$$

which we can use to evaluate the semiconductor capacitance,

$$C_s = \frac{q^2 N_{2d}}{k_b T} \left(\frac{1}{1 + e^{-\eta_c}} \right) = 23 \, \mu\text{F/cm}^2 \, .$$

The semiconductor capacitance is lower when we use Fermi-Dirac statistics for n_s, so the gate capacitance is lower too:

$$C_g = \frac{C_{\text{ox}} C_s}{C_{\text{ox}} + C_s} = 3.3 \, \mu\text{F/cm}^2 \, .$$

We conclude that the gate capacitance is roughly 15% lower than C_{ox}. The difference between C_g and C_{ox} must be accounted for when analyzing modern MOSFETs.

From Eq. (5.28) we see that if $C_{ox} \ll C_s$, then $C_g \approx C_{ox}$. This was the default assumption for decades, but as the gate oxide thickness was continually downscaled and then when SiO_2 was replaced by high-k gate dielectrics, C_{ox} became large enough that the effect of the semiconductor capacitance could no longer be ignored.

5.8 References

For a more complete discussion of the mobile charge in bulk MOSFETs and in MOSFETs with fully depleted thin channels, see Lectures 8 and 9 in:

[1] Mark Lundstrom, *Fundamentals of Nanotransistors*, World Scientific Publishing Co., Singapore, 2018.

For a review of concepts such as a particle in a box, 2D density-of-states, intrinsic carrier concentration, see:

[2] Robert F. Pierret *Advanced Semiconductor Fundamentals*, 2nd Ed., Vol. VI, Modular Series on Solid-State Devices, Prentice Hall, Upper Saddle River, N.J., USA, 2003.

An empirical model that describes $Q_n(V_{GS})$ from subthreshold to strong inversion has been presented by Wright.

[3] G.T. Wright "Threshold modelling of MOSFETs for CAD of CMOS VLSI," *Electron. Lett.* **21**, pp. 221–222, 1985.

The MIT Virtual Source Model is described in:

[4] A. Khakifirooz, O.M. Nayfeh, and D.A. Antoniadis, "A Simple Semiempirical Short-Channel MOSFET Current-Voltage Model Continuous Across All Regions of Operation and Employing Only Physical Parameters," *IEEE Trans. Electron. Dev.*, **56**, pp. 1674–1680, 2009.

Lecture 6

IV Theory: The Ballistic MOSFET

6.1 Introduction
6.2 Unidirectional thermal velocity
6.3 Ballistic IV characteristics
6.4 Linear region
6.5 Saturation region
6.6 Velocity at the virtual source
6.7 Summary
6.8 References

6.1 Introduction

Traditional MOSFET theory as summarized in Lecture 4 does not provide a good starting point for understanding nanoscale MOSFETs. A better starting point is the ballistic channel MOSFET in which charge carriers are injected into the channel from the source and flow across to the drain without scattering from lattice vibrations, defects, etc. It is important to understand that we are talking about ballistic transport across the channel; we assume that there is strong scattering in the much larger, heavily doped source and drain regions. Strong scattering maintains near-equilibrium conditions so that Fermi levels are well-defined in the source and drain. Modern III-V channel FETs operate very close to the ballistic limit, but for short channel Si MOSFETs, carrier scattering causes them to operate below the ballistic limit. In Lecture 7, we'll discuss how to extend the ballistic model to include carrier scattering. For a more extensive treatment of the ballistic MOSFET, including the use of Fermi-Dirac statistics for electrons, see [1].

Figure 6.1 shows $E_c(x)$ vs. x in the on-state with the source and drain Fermi levels indicated. As discussed in Lecture 3, the magnitude of the drain current is determined by the height of the energy barrier between the source and the channel. In an electrostatically well-designed MOSFET, the top of the barrier (the virtual source) is a low-field region strongly under the control of the gate potential. As indicated in Fig. 6.1, once $V_{DS} > V_{DSAT}$, additional increases in drain voltage have little effect on I_D because the barrier height changes very little (i.e. DIBL is low). The current will be evaluated at $x = 0$, the top of the barrier because at that point in a well-designed MOSFET the equilibrium $Q_n(V_{GS})$ relations developed in Lecture 5 apply. The drain current is

$$I_D = q\left(F^+ - F^-\right),\tag{6.1}$$

where F^+ is the magnitude of the electron flux injected from the source, and F^- is the magnitude of the flux injected from the drain. Before we compute $I_D(V_{GS}, V_{DS})$, we need to understand the average x-directed velocity of the charge carriers in these two streams.

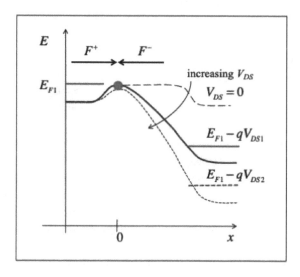

Fig. 6.1   Energy band diagram of a MOSFET under high gate bias for three different values of V_{DS}: i) $V_{DS} = 0$, ii) $V_{DS} = V_{DS1} > V_{DSAT}$, and iii) $V_{DS2} > V_{DS1}$. Also shown is the location the the virtual source at the top of the barrier and the two current streams in a ballistic MOSFET, F^+ and F^-.

6.2   Unidirectional thermal velocity

Consider electrons at an energy, E, in the conduction band. The probability that states at E are occupied is given by the equilibrium Fermi function,

$$f_0(E) = \frac{1}{1 + e^{(E-E_F)/k_BT}} \approx e^{-(E-E_F)/k_BT}$$

$$= e^{\left(E_F - E_c' - \hbar^2(k_x^2 + k_z^2)/2m^*\right)/k_BT} \propto e^{-m^*(v_x^2 + v_z^2)/2k_BT}, \tag{6.2}$$

which is a *Maxwellian distribution* of velocities in the $x - z$ plane of the channel as shown in Fig. 6.2a. (The $+x$ direction is along the channel, the width of the channel is along z, and y is normal to the plane of the channel.) We have assumed a nondegenerate semiconductor and a parabolic energy band for which the kinetic energy is $E - E_c' = \hbar^2(k_x^2 + k_z^2)/2m^* = m^*(v_x^2 + v_z^2)/2$. While the average velocity in any direction is zero in equilibrium, the average velocity of only those electrons with velocities in the $+x$-direction is a positive quantity equal to the magnitude of the average velocity of electrons with velocities only in the $-x$ direction. Figure 6.2b shows a hemi-Maxwellian velocity distribution with only $+x$-directed velocities.

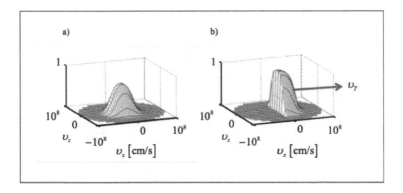

Fig. 6.2   Maxwellian distribution of velocities for electrons in the conduction band of a nondegenerate semiconductor. Left a): Distribution of velocities in the $x - z$ plane of the channel. Right b): Distribution of velocities with a component along the $+x$-axis. The careful observer will note that the hemi-Maxwellian distribution in Fig. 6b) is larger than the positive half of the equilibrium distribution in Fig. 6a). We'll explain why later in this lecture. (Reprinted from *Solid-State Electron.*, **46**, pp. 1899-1906, J.-H. Rhew, Zhibin Ren, and Mark Lundstrom, "A Numerical Study of Ballistic Transport in a Nanoscale MOSFET," ©2002, with permission from Elsevier.)

Figure 6.3 shows a velocity vector in the $x - z$ plane at a specific energy, E, and angle, θ, with the x-axis. The x-directed velocity is $v(E)\cos\theta$. The average velocity of electrons with a $+x$ velocity component is

$$\langle v_x^+(E) \rangle = \frac{\int_{-\pi/2}^{+\pi/2} v(E)\cos\theta\, d\theta}{\pi} = \frac{2}{\pi} v(E), \qquad (6.3)$$

where the brackets, $\langle \cdot \rangle$, denote an average over angle in the $x - z$ plane at a specific energy, E.

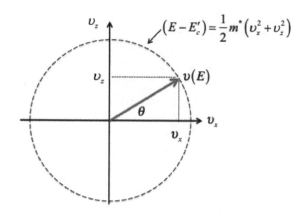

Fig. 6.3   An electron velocity vector at energy, E, in the $x - z$ plane. For a parabolic energy band, the magnitude of the velocity (length of the vector) is determined by the energy and is independent of direction. $\langle v_x^+(E) \rangle$ is the average x-directed velocity for $-\pi/2 < \theta < \pi/2$.

Electrons are distributed over a small range of energies near E_c' so Eq. (6.3) must be averaged over energy,

$$\langle v_x^+ \rangle = \frac{\int_{E_c'}^{\infty} \langle v_x^+(E) \rangle D_{2D}(E) f_0(E) dE}{\int_{E_c'}^{\infty} D_{2D}(E) f_0(E) dE}$$

$$\qquad\qquad (6.4)$$

$$= \frac{\int_{E_c'}^{\infty} \frac{2}{\pi} v(E) D_{2D}(E) f_0(E) dE}{\int_{E_c'}^{\infty} D_{2D}(E) f_0(E) dE}.$$

A word about notation. The quantity, $\langle v_x^+(E) \rangle$, denotes an average over angle at an energy, E, while the quantity, $\langle v_x^+ \rangle$, denotes an average over both angle and energy.

In general, the *unidirectional thermal velocity* depends on the location of the Fermi level, but in the nondegenerate limit, it is independent of Fermi level (see Exercise 12.2 in [1]):

$$\boxed{\langle v_x^+ \rangle = v_T = \sqrt{\frac{2k_B T}{\pi m^*}}} .$$
(6.5)

The unidirectional thermal velocity plays a key role in the theory of the ballistic MOSFET, so it is important to be calibrated on its magnitude. For electrons in Si, $v_T \approx 10^7$ cm/s, but with the lighter electron effective masses in III-V semiconductors, v_T can be a few times 10^7 cm/s. Under on-state conditions in Si MOSFETs, v_T can increase significantly due to carrier degeneracy. This occurs because the conduction band is filled to higher energies where the band velocity is larger. (See Sec. 14.4 in [1]). To keep the mathematics simple and the focus on essential concepts, we'll use the nondegenerate unidirectional thermal velocity in these lectures.

6.3   Ballistic IV characteristics

The *IV* characteristic of a ballistic MOSFET can be derived with a simple, *thermionic emission* model. As given by Eq. (6.1), the net drain current is q times F^+ (the flux of electrons from the source, over the barrier, and out the drain) minus F^- (the magnitude of the electron flux from the drain, over the barrier, and out the source). The probability that an electron in the source can surmount the energy barrier is $\exp(-E_{SB}/k_B T)$, where E_{SB} is the barrier height from the source to the top of the barrier, so F^+ is

$$F^+ \propto e^{-E_{SB}/k_B T} .$$
(6.6)

The probability that an electron from the drain can surmount the barrier is $\exp(-E_{DB}/k_B T)$, where E_{DB} is the barrier height from the drain to the top of the barrier, so F^- is

$$F^- \propto e^{-E_{DB}/k_B T} .$$
(6.7)

Because the drain voltage pulls the conduction band in the drain down, $E_{DB} > E_{SB}$. When there is no DIBL, $E_{DB} = E_{SB} + qV_{DS}$, so $F^+/F^- = \exp(-qV_{DS}/k_B T)$, and we can write the net drain current as

$$I_D = q\left(F^+ - F^-\right) = qF^+ \left(1 - e^{-qV_{DS}/k_B T}\right) .$$
(6.8)

At the top of the barrier, there are two streams of electrons, one moving to the right and one moving to the left. They have the same kinetic energy, so their velocities, v_T, are the same. The left to right flux is $qF^+ = W|Q_n^+(x = 0)|v_T$, where $Q_n^+(x = 0)$ is the charge in C/cm^2 at the top of the barrier due to electrons with positive velocities, and W is the width of the MOSFET. Similarly, $qF^- = W|Q_n^-(x = 0)|v_T$. We find the total charge by adding the charge in the two streams,

$$Q_n(x = 0) = -q\frac{(F^+ + F^-)}{Wv_T} = -q\frac{F^+}{Wv_T}\left(1 + F^+/F^-\right)$$
$$= -q\frac{F^+}{Wv_T}\left(1 + e^{-qV_{DS}/k_BT}\right). \tag{6.9}$$

Finally, if we solve Eq. (6.9) for F^+ and insert the result in Eq. (6.8), we find the *IV* characteristic of a ballistic MOSFET as

$$\boxed{I_D = W|Q_n(V_{GS}, V_{DS})|v_T\frac{(1 - e^{-qV_{DS}/k_BT})}{(1 + e^{-qV_{DS}/k_BT})}.} \tag{6.10}$$

6.4   Linear region

In this section and the next, we'll examine the general result, Eq. (6.10), under low and high drain bias. For small drain bias, we use $\exp(x) \approx 1 + x$ for small x to find

$$I_{DLIN} = W|Q_n(V_{GS})|\frac{v_T}{2k_BT/q}V_{DS}, \tag{6.11}$$

which is independent of channel length, L. (We have dropped the V_{DS} dependence of Q_n because for a well-designed transistor, the dependence on V_{DS} is small.) Equation (6.11) looks much different from the traditional expression, Eq. (4.5),

$$I_{DLIN} = \frac{W}{L}\mu_n|Q_n(V_{GS})|V_{DS}, \tag{6.12}$$

but we can make Eq. (6.11) look like the traditional model if we multiply and divide it by L to find

$$I_{DLIN} = \frac{W}{L}|Q_n(V_{GS})|\left(\frac{v_T L}{2(k_BT/q)}\right)V_{DS}. \tag{6.13}$$

Note that the dimensions of the quantity in parentheses are m^2/V $-$ s, the dimensions of mobility. Accordingly, we define the *ballistic mobility* as

$$\boxed{\mu_B \equiv \left(\frac{v_T L}{2(k_BT/q)}\right).} \tag{6.14}$$

Finally, we write the linear region current in the ballistic limit as

$$I_{DLIN} = \frac{W}{L}\mu_B |Q_n(V_{GS})| V_{DS}, \tag{6.15}$$

which is exactly the traditional expression for the linear region current except that the mobility has been replaced by the ballistic mobility. The question is: "What is the physical significance of the ballistic mobility?"

The transport of electrons subject to randomizing scattering events in a bulk semiconductor is described by the mobility, a material dependent parameter given by (see [1], Sec. 12.8):

$$\boxed{\mu_n \equiv \left(\frac{v_T \lambda}{2(k_B T/q)} \right)}, \tag{6.16}$$

where λ is the *mean-free-path for backscattering*, roughly speaking, the average distance between scattering events. (See [1], Sec. 12.4 for the specific definition). In a ballistic channel MOSFET, there is no scattering in the channel, but electrons in the source and drain contacts scatter frequently, so the distance between scattering events is the length of the channel. It seems sensible, therefore, to replace the actual mean-free-path by the length of the channel, and that leads to the concept of a ballistic mobility. The ballistic mobility is a way to write I_{DLIN} of a ballistic MOSFET in the traditional form, but it also has a clear, physical interpretation.

To calibrate ourselves, we can estimate μ_B for a nanoscale Si MOSFET with $L = 10$ nm. For electrons in Si, $v_T \approx 10^7$ cm/s and at 300 K, $k_B T/q = 0.026$ V, so $\mu_B \approx 200 \, \text{cm}^2/\text{V} - \text{s}$, which is also about what the scattering limited mobility, μ_n is.

6.5   Saturation region

Consider next the high V_{DS}, saturation current, I_{DSAT}. In this case, $F^- \ll F^+$, and the drain current saturates at $I_D = qF^+$. In the limit, $V_{DS} \gg k_B T/q$, Eq. (6.10) becomes

$$I_{DSAT} = W|Q_n(V_{GS}, V_{DS})|v_{inj}, \tag{6.17}$$

where v_{inj}, the so-called *ballistic injection velocity*,

$$v_{inj} = v_T = \sqrt{\frac{2k_B T}{\pi m^*}}, \tag{6.18}$$

is the unidirectional thermal velocity. The injection velocity is the thermal average velocity at which electrons are injected from the source to the virtual source at the top of the barrier. For $V_{DS} > V_{DSAT}$, the drain current increases slowly because DIBL causes Q_n to increase with drain voltage.

Equation (6.17), the saturated region current for a ballistic MOSFET, looks almost identical to the traditional velocity saturation expression, Eq. (4.7),

$$I_{DSAT} = W|Q_n(V_{GS}, V_{DS})|v_{sat}. \tag{6.19}$$

The ballistic and traditional models not only look similar, they are also numerically similar for Si MOSFETs because $v_T \approx v_{sat} \approx 10^7$ cm/s, but for III-V FETs, $v_T > v_{sat}$. It is curious that the ballistic and traditional expressions are so similar because in the traditional model, velocity saturation is the result of strong scattering in the high electric field near the drain. How does the velocity saturate when there is no scattering?

6.6   Velocity at the virtual source

As given by Eq. (4.1), the drain current is the product of charge and velocity at the top of the barrier. By equating Eq. (4.1) to Eq. (6.10), I_D for the ballistic MOSFET,

$$\begin{aligned}
I_D &= W|Q_n\,(V_{GS})|\,\langle v_x \rangle \\
&= W|Q_n(V_{GS})|v_T \frac{\left(1 - e^{-qV_{DS}/k_BT}\right)}{\left(1 + e^{-qV_{DS}/k_BT}\right)},
\end{aligned}$$

we find the average velocity at the virtual source to be

$$\begin{aligned}
\langle v_x \rangle &= v_T \left[\frac{1 - e^{-qV_{DS}/k_BT}}{1 + e^{-qV_{DS}/k_BT}}\right] \\
\langle v_x^+ \rangle &= v_T = \sqrt{\frac{2k_BT}{\pi m^*}}.
\end{aligned} \tag{6.20}$$

(It is important to understand that $\langle v_x \rangle$ is the average velocity in the x-direction of *all* electrons while $\langle v_x^+ \rangle = v_T$ is the average velocity in the x-direction of *only* the electrons with a $+x$ velocity component.) Figure

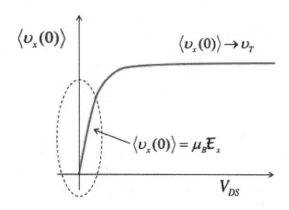

Fig. 6.4 Sketch of the average velocity at the virtual source vs. V_{DS}. Maxwell-Boltzmann statistics are assumed.

6.4 is a sketch of $\langle v_x(x = 0) \rangle$ vs. V_{DS}. For low V_{DS}, $\langle v_x(0) \rangle \propto V_{DS}$, and for high V_{DS}, $\langle v_x(0) \rangle$ saturates at v_T.

The velocity vs. drain voltage sketched in Fig. 6.4 is much like the velocity vs. electric field characteristic in a bulk semiconductor; the velocity is proportional to the drain voltage for low voltage, and it saturates at high voltages. Note, however, that this velocity is at the top of the source to channel barrier, at $x = 0$. The velocity saturates at the source end of the channel, at the top of the barrier, where the electric field is zero and not at the drain end of the channel where the electric field is high.

To examine $\langle v_x(0) \rangle$ for small V_{DS}, we expand the exponentials in Eq. (6.20) for small argument ($\exp(x) \approx 1 + x$) to find

$$\langle v_x(0) \rangle = \frac{v_T}{2k_BT/q}V_{DS}, \qquad (6.21)$$

and then multiply and divide by the channel length, L, to find

$$\langle v_x(0) \rangle = \left(\frac{v_T L}{2k_BT/q} \right) \frac{V_{DS}}{L}. \qquad (6.22)$$

The first term on the RHS can be recognized as the ballistic mobility, and the second term is the electric field in the channel, so the velocity for low V_{DS} can be written as

$$\langle v_x(0) \rangle = \mu_B \mathcal{E}_x. \qquad (6.23)$$

In a bulk semiconductor, $\langle v_x \rangle = \mu_n \mathcal{E}_x$ with μ_n being the scattering limited mobility as given by Eq. (6.16). The low V_{DS} velocity at the top of the barrier in the ballistic MOSFET can be written similarly, if we replace μ_n with μ_B.

Equation (6.20) and Fig. 6.4 show that in a ballistic MOSFET, the average velocity at the virtual source saturates when V_{DS} is greater than a few $k_B T/q$. To understand how this occurs, we should examine the distribution of velocities in the $x - z$ plane of the channel. Recall that in a nondegenerate, bulk semiconductor in equilibrium, the velocities are distributed in a Maxwellian distribution as shown in Fig. 6.2a. A ballistic MOSFET under high drain bias is far from equilibrium, so we expect the distribution of carrier velocities to be much different from the equilibrium distribution shown in Fig. 6.2a.

Figure 6.5 shows the results of a numerical simulation for a 10 nm channel length ballistic MOSFET. The gate voltage is high, so the source to channel energy barrier is low. As indicated in Fig. 6.5a, we seek to understand the distribution of carrier velocities at the top of the barrier as the drain voltage increases from $V_{DS} = 0$ V to $V_{DS} = V_{DD}$.

In a ballistic MOSFET, the distribution of velocities at the top of the barrier consists of two components, a positive velocity component injected from the source and a negative velocity component injected from the drain. These two components are given by

$$f^+(v_x > 0, v_z) = e^{(E_{FS} - E'_c(0))/k_B T} \times e^{-m^*(v_x^2 + v_z^2)/2k_B T}$$
$$f^-(v_x < 0, v_z) = e^{(E_{FD} - E'_c(0))/k_B T} \times e^{-m^*(v_x^2 + v_z^2)/2k_B T},$$

(6.24)

where E_{FS} is the Fermi level in the source, and $E_{FD} = E_{FS} - qV_{DS}$ is the Fermi level in the drain. As V_{DS} increases, the magnitude of $f^-(v_x, v_z)$ decreases.

Figure 6.5b is a plot of the velocity distributions at four different drain voltages. Consider first case i), $V_{DS} = 0$, where the velocity distribution has an equilibrium shape. Since $V_{DS} = 0$, no current flows, and the MOSFET is in equilibrium, so the observation of an equilibrium distribution of velocities is not a surprise, but there is no scattering in a ballistic MOSFET, so how is equilibrium established? The answer is that at the top of the barrier, all electrons with $v_x > 0$ came from the source, in which strong electron-phonon scattering maintains equilibrium. Also at the top of the barrier, all electrons with $v_x < 0$ came from the drain, where strong electron-phonon scattering maintains equilibrium. Since $E_{FS} = E_{FD}$ at $V_{DS} = 0$, the magnitudes of the positive and negative components are equal, so an overall

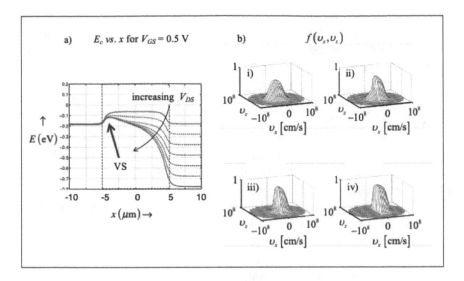

Fig. 6.5   Results of numerical simulations of a ballistic MOSFET. Left a): $E_c(x)$ vs. x at a high gate voltage for various drain voltages. Right b): The velocity distributions at the top of the barrier at four different drain voltages: i) 0 V, ii) 0.05 V, iii) 0.1 V, and iv) 0.6 V. (Reprinted from *Solid-State Electron.*, **46**, pp. 1899-1906, J.-H. Rhew, Zhibin Ren, and Mark Lundstrom, "A Numerical Study of Ballistic Transport in a Nanoscale MOSFET," ©2002, with permission from Elsevier.)

equilibrium Maxwellian velocity distribution results. Even though there is no scattering near the top of the barrier, the distribution of velocities is an equilibrium one.

Consider next case ii), $V_{DS} = 0.05$ V. The magnitude of the negative velocity component is smaller, so there are fewer negative velocity electrons but nearly the same number of positive velocity electrons, so the average x-directed velocity is positive. For this small V_{DS} regime, the average velocity increases linearly with V_{DS}. The case iii), $V_{DS} = 0.1$ V velocity distribution, shows an even smaller negative velocity component. Finally, case iv), the $V_{DS} = 0.6$ V velocity distribution, shows no negative velocity electrons because the drain Fermi level has been lowered so much that the probability of a negative velocity state at the top of the barrier being occupied is negligibly small. The average x-directed velocity is as large as it can be; further increases in the drain voltage will not increase the velocity — the velocity has saturated.

Figure 6.5 explains the velocity vs. drain voltage characteristic we derived in Eq. (6.20), but there is a subtle point that should be discussed. A careful look at the hemi-Maxwellian distribution for $V_{DS} = 0.6$ V shows that it is larger than the positive half of the equilibrium distribution for $V_{DS} = 0$ V. A careful look at the left figure in Fig. 6.5 shows that $E'_c(0)$ is pushed down for increasing V_{DS}. This is not DIBL; it is a result of MOS electrostatics in a well-designed MOSFET.

In a well-designed MOSFET, the charge at the top of the barrier, $Q_n(0)$, depends mostly on the gate voltage and does not change substantially with increasing drain voltage (i.e. the DIBL is low). As the population of negative velocity electrons decreases with increasing V_{DS}, more positive velocity electrons must be injected to balance the charge on the gate. Since the source Fermi level does not change, Eq. (6.24) shows that $E'_c(0)$ must decrease in order to increase the charge injected from the source and satisfy MOS electrostatics.

Finally, we note that the overall shapes of the velocity distributions for $V_{DS} > 0$ are much different from the equilibrium shape, but each half has an equilibrium shape. Scattering is what returns a system to equilibrium, and there is no scattering in the channel of a ballistic MOSFET. The ballistic device is very far from equilibrium, but each half of the velocity distribution is in equilibrium with one of the two contacts.

6.7 Summary

In this lecture, a simple theory of the ballistic MOSFET was discussed. The important results are summarized in the following four equations.

$$I_{DLIN} = \frac{W}{L}|Q_n(V_{GS})|\mu_B V_{DS} \tag{6.15}$$

$$\mu_B \equiv \left(\frac{v_T L}{2(k_B T/q)}\right) \tag{6.14}$$

$$I_{DSAT} = W|Q_n(V_{GS}, V_{DS})|v_{inj} \tag{6.17}$$

$$v_{inj} = v_T = \sqrt{\frac{2k_B T}{\pi m^*}} \tag{6.18}$$

According to Eq. (6.15), I_{DLIN} for a ballistic MOSFET is given by the traditional expression except that the mobility, μ_n is replaced by the ballistic mobility, μ_B. Equation (6.14) shows that the ballistic mobility is just the scattering limited mobility with the mean-free-path for backscattertng replaced by the channel length, L. According to Eq. (6.17), I_{DSAT} for a balllistic MOSFET is given by the traditional velocity saturation model with v_{sat} replaced by the ballistic injection velocity. Equation (6.18) shows that the ballistic injection velocity is just the thermal equilibrium unidirectional thermal velocity, v_T.

When $V_{GS} > V_T$, $|Q_n| = C_g (V_{GS} - V_T)$, so we see that I_{DSAT} increases linearly with $(V_{GS} - V_T)$, which is the signature of velocity saturation in the IV characteristic. We also discussed in this lecture how the velocity saturates in a ballistic MOSFET. It does not saturate due to scattering in the high field region near the drain. Instead, it saturates when $F^- \ll F^+$ where qF^+ is the ballistic current injected from the source and qF^- is the ballistic current injected from the drain.

Let's estimate the ballistic on-current in a Si MOSFET. Under on-state conditions, $n_S \approx 10^{13} \, \text{cm}^{-2}$ and $v_T \approx 10^7 \, \text{cm/s}$, so for $W = 1 \, \mu\text{m}$, we find

$$I_{on} = W|Q_n|v_T = 10^{-4} \times (1.6 \times 10^{-19}) \times 10^{13} \times 10^7 = 1.6 \quad \text{mA} \,,$$

which is well above the 1 mA/μm on-current for a typical, n-channel Si MOSFET. The actual current is below the ballistic limit because of carrier scattering — not all of the electrons injected into the channel from the source make it across to the drain. Extending the ballistic model to include carrier scattering is the subject of the next lecture.

Lecture 6 Exercise: Ballistic channel resistance

The channel resistance, R_{ch}, is obtained from Eq. (6.11), the ballistic expression for I_{DLIN}, as

$$I_{DLIN} = W|Q_n(V_{GS})| \left(\frac{v_T}{2k_BT/q} \right) V_{DS} = V_{DS}/R_{ch}^{ball}$$

$$R_{ch}^{ball} = \frac{1}{W\left[v_T/(2k_BT/q)\right]|Q_n(V_{GS})|} \, ,$$

which is independent of channel length as expected in the ballistic limit. From Eq. (4.5), the traditional expression for I_{DLIN}, we can find the diffusive channel resistance as

$$I_{DLIN} = \frac{W}{L}\mu_n|Q_n(V_{GS})|V_{DS}$$

$$R_{ch}^{diff} = \frac{L}{W\mu_n\,|Q_n(V_{GS})|} \, ,$$

which is proportional to the channel length, L, as expected.

In this exercise, we will compare the ballistic and diffusive channel resistances. Assume a 10 nm channel length MOSFET with $V_{GS} = V_{DD}$ so that $n_s \approx 10^{13}\,\text{cm}^{-2}$. Assume a scattering limited mobility of $\mu_n = 250\,\text{cm}^2/\text{V} - \text{s}$, and $T = 300$ K. Assume $W = 1\,\mu\text{m}$, so that the resistances can be expressed in $\Omega - \mu\text{m}$.

The mobile charge density is

$$|Q_n| = qn_s = 1.6 \times 10^{-6} \text{ C/cm}^2 \, .$$

For the uni-directional thermal velocity, we assume $m^* = 0.19m_0$, the transverse electron mass in the Si conduction band, which is appropriate for the lowest subband in (100) Si (see Lecture 9 in [1]). We find

$$v_T = \sqrt{\frac{2k_BT}{\pi m^*}} = 1.2 \times 10^7 \text{ cm/s} \, .$$

Computing the ballistic channel resistance, we find

$$R_{ch}^{ball} = \frac{1}{W\left[v_T/(2k_BT/q)\right]|Q_n(V_{GS})|} = 27\,\Omega - \mu\text{m} \, .$$

Computing the diffusive channel resistance, we find

$$R_{ch}^{diff} = \frac{L}{W\mu_n\,|Q_n(V_{GS})|} = 25\,\Omega - \mu\text{m} \, .$$

We see that $R_{ch}^{ball} \approx R_{ch}^{diff}$, and we will see in the next lecture, that the actual channel resistance is the sum of the ballistic and diffusive channel resistances. Because the two resistances are approximately equal, this exercise shows that modern Si MOSFETs operate squarely in the quasi-ballistic transport regime — between the diffusive and ballistic limits.

6.8   References

For a more complete discussion of the ballistic MOSFET including the use of Fermi-Dirac statistics and quantum confinement, see Lectures 12–15 in:

[1] Mark Lundstrom, *Fundamentals of Nanotransistors*, World Scientific Publishing Co., Singapore, 2018.

IV Theory: Transmission Approach

7.1   Introduction

Having discussed ballistic MOSFETs, we will now examine how carrier scattering affects the IV characteristics of nanoscale MOSFETs. Although scattering complicates things, the basic principles are readily understood and can be translated into a simple model for the IV characteristics.

Figure 7.1 compares a carrier trajectory in a ballistic channel MOSFET to one in which scattering occurs. As shown in Fig. 7.1a for a ballistic MOSFET, electrons are injected from the source (where they scatter frequently) into the channel (where they don't scatter at all) and then exit by entering the drain (where they again scatter frequently). The potential drop in the channel accelerates electrons, so they gain kinetic energy. The kinetic energy is deposited in the drain.

Figure 7.1b shows one possible carrier trajectory in the presence of scattering. Note that some scattering events are *elastic*, the carrier changes direction but the energy does not change.   Some scattering events are *inelastic* — both the direction and energy of the electron change.   For example, electrons can gain energy by absorbing a lattice vibration (a *phonon*), and they can lose energy by exciting a lattice vibration (generating a phonon). For the particular trajectory shown, the electron injected from the source exits through the drain, but scattering is a stochastic process, and for some carrier trajectories, electrons injected from the source backscatter and return to the source. The *transmission*, $\mathcal{T}$, is the ratio of the flux of electrons injected from the source to the flux that exits at the drain; carrier scattering reduces the transmission.

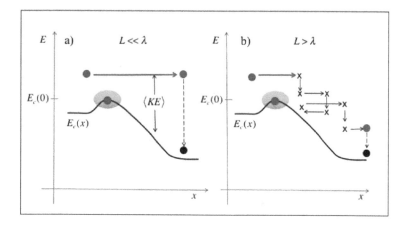

Fig. 7.1 Ballistic vs. diffusive transport in a MOSFET. Left a): Trajectory of an electron injected from the source into a ballistic channel. Right b): Trajectory of an electron injected from the source into a channel in which the electron scatters several times. Scattering is a stochastic process, so this trajectory is just one of a large ensemble of possible trajectories. The shaded region near the virtual source is under the strong control of the gate, and the electric field along the channel, $\mathcal{E}_x$ is low in this region.

Nanoscale MOSFETs are neither fully ballistic ($\mathcal{T} = 1$) nor fully diffusive ($\mathcal{T} \ll 1$); they typically operate in a *quasi-ballistic* regime where $\mathcal{T} \lesssim 1$. Using the concept of transmission, the results for the ballistic MOSFET are readily extended to include carrier scattering. For more on the physics of carrier scattering and transmission, see Lecture 16 in [1] or Lecture 6 in [2]. Before we discuss the *IV* characteristics, however, we need to understand the physics of transmission.

7.2 Transmission

i) Transmission when the electric field is small

Consider the problem illustrated in Fig. 7.2, which assumes 2D electrons in the $x - z$ plane of the channel. A steady-state flux of electrons, $F^+(x = 0)$, is injected from the source into a channel of length, L; there is no electric field in the channel. A flux, $F^+(x = L)$ emerges from the right and enters the drain. For the case shown in Fig. 7.2, no flux is injected from the right (drain). Within the channel, backscattering reverses the direction of fluxes, so there is both an $F^+(x)$ and an $F^-(x)$ for $0 < x < L$. We assume near equilibrium conditions, so $F^+(x) = n_s^+(x)v_T$ m^{-1}s^{-1}, where n_s^+ m^{-2} is the concentration of 2D electrons with velocities in the $+x$ direction. For small V_{DS}, electrons are injected from both the source and drain; we could solve the two cases independently and add the results, but in the absence of a significant electric field in the channel, the problem is symmetric, so we will only consider the source-injected flux shown in Fig. 7.2.

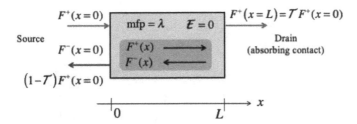

Fig. 7.2 Illustration of how transmission, $\mathcal{T}$, is defined. An equilibrium flux of electrons is injected from the source, and the drain is treated as an "absorbing contact." No flux is injected from the drain, and any flux that enters the drain is absorbed and not reflected back into the channel. There is an analogous case for a flux injected from the drain.

The transmission is the ratio of the flux emerging at $x = L$ to the incident flux at $x = 0$,

$$\mathcal{T} \equiv \frac{F^+(x = L)}{F^+(x = 0)}, \tag{7.1}$$

which is a number between zero and one. Some part of the injected flux transmits across the channel and enters the drain, and some part, $F^-(x = 0)$, backscatters and returns to the source. Assuming that there is no recombination or generation in the slab, then $F^+(0) = F^-(0) + F^+(L)$, from which we find that $F^-(0) = (1 - \mathcal{T})F^+(0)$. We assume that the transmission from the source to drain as shown in Fig. 7.2 is identical to the transmission from the drain to source because under low V_{DS}, the channel is symmetric.

It is clear that the transmission will depend on the mean-free-path (mfp) for backscattering, λ, and on the length of the channel. Consider first a very short channel for which $L \ll \lambda$. In this case, almost all of the injected flux emerges at the right. There is no backscattered flux, so $F^+(L) = F^+(0)$ and $F^-(0) = 0$. In this *ballistic limit*, the transmission is one,

$$\mathcal{T}_{ball} = 1 \,. \tag{7.2}$$

Next, consider the *diffusive limit* for which $L \gg \lambda$. The channel is many mean-free-paths long, so we expect the transmission to be small, $\mathcal{T}_{diff} \ll 1$. This is the case for micrometer long channels. To compute the transmission in the diffusive limit, let's first compute $n_s(0)$ and $n_s(L)$. At $x = 0$, n_s consists of electrons with both positive and negative x-directed velocities. We find

$$\begin{aligned}
n_s(0) &= n_s^+(0) + n_s^-(0) \\
&= F^+(0)/v_T + F^-(0)/v_T \\
&= \left(F^+(0)/v_T\right)\left[1 + (1 - \mathcal{T}_{diff})\right] \approx 2F^+(0)/v_T \,,
\end{aligned}$$

where the last result follows from the fact that $\mathcal{T}_{diff} \ll 1$. At $x = 0$, $n_s^-(0) \approx n_s^+(0)$; there is a near-equilibrium, symmetrical distribution of velocities. The positive half comes from the equilibrium flux injected from the source, and because $L \gg \lambda$, most of the injected flux backscatters and produces a near-equilibrium negative half.

Next, let's compute $n_s(L)$. At $x = L$, there are no negative velocity electrons because none are injected from the drain. In this example, we find

$$\begin{aligned}
n_s(L) &= n_s^+(L) + n_s^-(L) \\
&= n_s^+(L) \\
&= \mathcal{T}_{diff}\left(F^+(0)/v_T\right) \approx 0,
\end{aligned}$$

where the last result follows from the fact that $\mathcal{T}_{diff} \ll 1$. Finally, the net flux in the diffusive limit is

$$F = F^+ - F^-$$
$$= F^+(0) - F^-(0) = F^+(L) = \mathcal{T}_{diff}F^+(0),$$

where we have assumed that there is no recombination or generation of carriers, so the net flux is independent of position.

In the diffusive limit, we also know that the net flux is also given by *Fick's Law* as

$$F = -D_n\frac{dn_s}{dx}, \tag{7.3}$$

where D_n is the diffusion coefficient. Diffusion is random thermal motion in the presence of scattering. It can be shown that D_n is related to the unidirectional thermal velocity and mfp for backscattering according to (Sec. 6.5 of [2]):

$$\boxed{D_n = \frac{v_T\lambda}{2} \quad \text{m}^2/\text{s}}, \tag{7.4}$$

which is an important result. Given the mobility, which is readily measured, we can use the Einstein relation, $D_n/\mu_n = k_BT/q$, to find D_n from which we can deduce the mean-free-path for backscattering.

Beginning from Fick's Law and using $n_s(0) = 2F^+(0)$, $n_s(L) = 0$, and $F = \mathcal{T}_{diff}F^+(0)$, we find

$$F = D_n\frac{[n_s(0) - n_s(L)]}{L}$$
$$= \frac{2D_nF^+(0)}{v_TL} = \mathcal{T}_{diff}F^+(0),$$

from which we can use Eq. (7.4) to find

$$\mathcal{T}_{diff} = \frac{\lambda}{L}.$$

(The assumed linearity of $n(x)$ can be proved by solving the equations for $F^+(x)$ and $F^-(x)$. See Sec. 6.3 in [2].)

We have derived the transmission in the ballistic and diffusive limits, but modern devices often operate in the *quasi-ballistic* regime between these two limits. The simplest expression for the transmission that gives the correct ballistic and diffusive limits is

$$\boxed{\mathcal{T} = \frac{\lambda}{\lambda + L}}. \tag{7.5}$$

Equation (7.5) plays an important role in the transmission theory of the MOSFET. It is clearly correct in the ballistic and diffusive limits, but it is also approximately correct between these limits. It can be derived from a simple Boltzmann Transport Equation (See Sec. 6.3 of [2]).

ii) Transmission when the electric field is large

Consider next the case where there is a strong electric field for $0 < x < L$. Now there are two transmissions, $\mathcal{T}_{LR}$ from left to right (from source to drain), and $\mathcal{T}_{RL}$ from right to left (from drain to source). Figure 7.3 is an energy band diagram showing carriers injected into a region with a strong electric field. An equilibrium flux of electrons, $F^+(0)$, is injected from the left (the source). The injected carriers quickly gain kinetic energy, and their scattering rate increases. Numerical simulations show that if the injected carriers penetrate just a short distance into the high field region without scattering, then even if they do subsequently scatter, they are bound to emerge from the right. Even when there is a significant amount of scattering, $\mathcal{T}_{LR} \approx 1$ because the high electric field sweeps carriers across and out the right contact. The high field region acts as a nearly perfect carrier collector — the absorbing contact shown in Fig. 7.2. An equilibrium flux of electrons, $F^-(L)$, is also injected from the right (from the drain towards the source), but electrons injected from the drain see a large potential barrier and are unlikely to transmit across to the source, so $\mathcal{T}_{RL} \approx 0$.

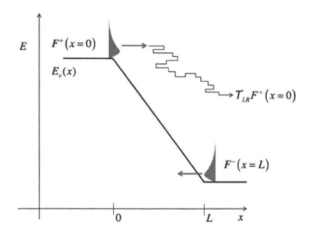

Fig. 7.3 One possible electron trajectory across a short region with a high electric field. Most electrons injected from an equilibrium distribution at the left exit at the right even if they scatter several times within the region. Electrons are also injected from the right, but they see a high potential barrier and cannot transmit across.

iii) Transmission: Key points

The simple picture discussed in this section summarizes insights gleaned from detailed simulations. The key points are as follows.

(1) Transmission is related to the mean-free-path for backscattering according to $\mathcal{T} = \lambda/(\lambda + L)$.
(2) For ballistic transport, $\mathcal{T} \to 1$, which happens when $L \ll \lambda$.
(3) For diffusive transport, $\mathcal{T} \to \lambda/L \ll 1$, which happens when $L \gg \lambda$.
(4) Regions with a high electric field are good carrier collectors, $\mathcal{T} \approx 1$.

For MOSFETs under low V_{DS}, the electric field in the channel is small, and the transmission is given by Eq. (7.5). For an electrostatically well-designed MOSFET under high V_{DS}, there is a short, low-field section near the beginning of the channel followed by a high-field section (as in Fig. 7.1) The transmission across the low-field section is given by Eq. (7.5) with L replaced by the length of the low-field section. The transmission across the high-field part of the channel is approximately one. Because the transmission across the entire channel is controlled by the short, low-field section near the virtual source, this part of the channel is a "bottleneck" that limits I_D.

Finally, we should mention that the mean-free-path, λ, is a specially defined "mean-free-path for backscattering". Physically, it is the average distance that a forward flux travels before it is backscattered to a negatively directed flux. More commonly, the mean-free-path is simply taken to be the average distance between scattering events. The mfp for backscattering is a clearly defined quantity that is always longer than the mfp for scattering. See Sec. 6.4 of reference [2] for a discussion.

Understanding the channel transmission coefficient under low and high V_{DS} involves some complex transport physics. Interested readers can find a longer discussion in Lecture 6 of [2] and the references cited therein, but with an understanding of the key points summarized above, we can readily compute I_{DLIN} and I_{DSAT}.

7.3   Linear region

In Lecture 6 (Eq. (6.15)), we developed a ballistic model for the linear region current,

$$I_{DLIN} = \frac{W}{L}\mu_B|Q_n(V_{GS})|V_{DS}, \qquad (7.6)$$

where μ_B is the ballistic mobility. In the presence of scattering, I_{DLIN} becomes

$$I_{DLIN} = \frac{W}{L}\left(\mathcal{T}_{lin}\,\mu_B\right)|Q_n(V_{GS})|V_{DS}, \qquad (7.7)$$

where $\mathcal{T}_{lin}$ is given by Eq. (7.5), but if we multiply the numerator and denominator of Eq. (7.5) by $v_T/(2(k_BT/q))$, we can express $\mathcal{T}_{lin}$ as

$$\mathcal{T}_{lin} = \frac{\mu_n}{\mu_n + \mu_B}, \qquad (7.8)$$

which allows us to write I_{DLIN} as

$$\boxed{I_{DLIN} = \frac{W}{L}\mu_{app}|Q_n(V_{GS})|V_{DS}}, \qquad (7.9)$$

where the *apparent mobility* is defined as

$$\boxed{\frac{1}{\mu_{app}} \equiv \frac{1}{\mu_n} + \frac{1}{\mu_B}}. \qquad (7.10)$$

To find the apparent mobility, we add the inverse mobility due to scattering to the inverse mobility due to ballistic transport and take the inverse of the sum. This prescription for finding the total mobility due to two independent processes is known as *Matthiessen's Rule* [3].

According to Eq. (7.10), the apparent mobility of a MOSFET is less than the lower of the scattering limited and ballistic mobilities. Recall that $\mu_B \propto L$. For a long channel MOSFET, $\mu_B \gg \mu_n$, and the apparent mobility is μ_n. For a very short channel, $\mu_B \ll \mu_n$, and the apparent mobility is μ_B. Note that the traditional expression for the linear current, Eq. (4.5), could predict a current above the ballistic limit if the channel length is short enough, but if the scattering limited mobility is replaced by the apparent mobility, this cannot happen.

In the linear region, the MOSFET is a gate-voltage controlled resistor with a channel resistance is obtained from Eq. (7.9) as

$$R_{ch} = \frac{V_{DS}}{I_{DLIN}} = \frac{L}{W}\frac{1}{|Q_n|\mu_{app}}. \qquad (7.11)$$

Note that there is a finite channel resistance even for a ballistic channel! Real MOSFETs also have series resistance, so in the linear region

$$I_{DLIN} = \frac{V_{DS}}{R_{ch} + R_S + R_D} = \frac{V_{DS}}{R_{tot}}, \qquad (7.12)$$

where R_S and R_D are the source and drain series resistances.

It is common to analyze MOSFET IV data to extract the mobility, but what is extracted is the apparent mobility. For a long channel MOSFET, the extracted mobility is the scattering limited mobility, μ_n, but for modern, short channel Si MOSFETs, $\mu_B \approx \mu_n$, so $\mu_{app} \approx \mu_n/2$. For III-V FETs, $\mu_n \gg \mu_B$, so μ_{app} is close to μ_B. When analyzing measured IV characteristics, it is the apparent mobility that determines the channel resistance. To summarize, we have shown that the transmission expression for the linear region current, Eq. (7.7), can be written in the traditional diffusive form, Eq. (4.5) — if we replace the scattering limited mobility, μ_n, by the apparent mobility, μ_{app}, as defined by Eq. (7.10).

7.4   Saturation region

Equation (6.17) In Lecture 6 describes I_{DSAT} for a ballistic MOSFET:

$$I_{DSAT} = W|Q_n(V_{GS}, V_{DS})|v_{inj}^{ball}, \qquad (7.13)$$

where v_{inj}, the ballistic injection velocity, is the unidirectional thermal velocity

$$v_{inj}^{ball} = v_T = \sqrt{\frac{2k_B T}{\pi m^*}}. \qquad (7.14)$$

In the presence of scattering, we might be tempted to just multiply the ballistic saturation current by the transmission as we did for I_{DLIN}, but the high drain bias case requires some care. Recall that the distribution of velocities at the top of the barrier for the ballistic MOSFET under high drain bias is a hemi-Maxwellian as shown in Fig. 6.5. In the presence of backscattering, there will be a negative velocity component to the velocity distribution. We have to ensure, however, that the total density of electrons, $n_s(x = 0)$, satisfies the requirements of MOS electrostatics, i.e. $qn_s(x = 0) = C_g (V_{GS} - V_T)$.

Consider the high drain bias case where injection from the drain to the top of the barrier is negligible. The ballistic case shown in Fig. 7.4a. A flux of electrons, F_{ball}^+, is injected from the source. In this case (high drain bias, ballistic transport), the only charge at the top of the barrier is charge injected from the source, $n_s(x=0) = n_s^+(x=0)$. Since current is charge times velocity, the charge at the top of the barrier is

$$Q_n(x=0) = -q n_s(x=0) = -q \frac{F_{ball}^+}{\upsilon_T}. \qquad (7.15)$$

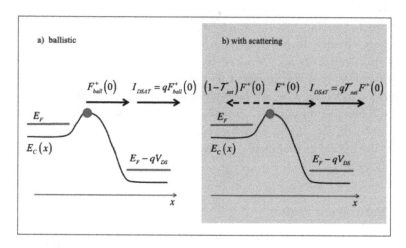

Fig. 7.4   Injected, transmitted, and backscattered currents under high drain bias. Left a): The ballistic case. Right b): In the presence of backscattering.

Next, consider the charge in the presence of scattering. As shown in Fig. 7.4b, there are two components of the charge: a source-injected component with positive velocity electrons, and a backscattered component with negative velocity electrons. The total charge is

$$Q_n(x=0) = -q \left[\frac{F^+ + (1 - \mathcal{T}_{sat})F^+}{\upsilon_T} \right] = -q \left[\frac{(2 - \mathcal{T}_{sat})F^+}{\upsilon_T} \right]. \qquad (7.16)$$

In a well-designed MOSFET, $Q_n(x=0)$ is largely determined by MOS electrostatics and is relatively independent of transport. The charge under ballistic conditions, Eq. (7.15) should be the same as the charge in the presence of scattering, Eq. (7.16). Equating Eq. (7.15) to (7.16), we find

$$F^+ = \frac{F_{ball}^+}{(2 - \mathcal{T}_{sat})}. \qquad (7.17)$$

In the ballistic case, $\mathcal{T}_{sat} = 1$ and $F^+ = F_{ball}^+$, but in the presence of scattering ($\mathcal{T} < 1$), so a smaller injected flux produces the same $Q_n(x=0)$.

The drain current is $\mathcal{T}_{sat}$ times the injected current, so from Eq. (7.17), we find the saturated drain current in the presence of scattering as

$$
\begin{aligned}
I_{DSAT} &= \mathcal{T}_{sat}(qF^+) \\
&= \mathcal{T}_{sat}\frac{qF^+_{ball}}{(2 - \mathcal{T}_{sat})} \\
&= \frac{\mathcal{T}_{sat}}{(2 - \mathcal{T}_{sat})}I^{ball}_{DSAT}.
\end{aligned}
\tag{7.18}
$$

The requirement that MOS electrostatics be enforced results in a saturation current that is $\mathcal{T}/(2 - \mathcal{T})$ times the ballistic saturation current — not $\mathcal{T}$ times the ballistic saturation current as might have been expected. Equation (7.13) in still valid in the presence of scattering, but the injection velocity changes:

$$
\boxed{
\begin{aligned}
I_{DSAT} &= W|Q_n(V_{GS}, V_{DS})|v_{inj} \\
v_{inj} &= \left(\frac{\mathcal{T}_{sat}}{2 - \mathcal{T}_{sat}}\right)v_T \leq v^{ball}_{inj} = v_T
\end{aligned}
}
\tag{7.19}
$$

We have been careful to distinguish between the transmission for low V_{DS}, $\mathcal{T}_{lin}$ and the transmission for high V_{DS}, $\mathcal{T}_{sat}$. Figure 7.5 shows why the two are different. Under low bias, the electric field is small across the entire channel as shown in Fig. 7.6a,. As discussed in Sec. 7.2, the transmission is determined by the length of the low-field region, so for low bias,

$$
\mathcal{T}_{lin} = \frac{\lambda_{lin}}{\lambda_{lin} + L}.
\tag{7.20}
$$

As shown in Fig. 7.5b, for high drain bias in a well-designed MOSFET, the low-field region is confined to a short section of length, ℓ, near the beginning of the channel. As discussed in Sec. 7.2, the high-field part of the channel acts as a near-perfect carrier collector with $\mathcal{T} \approx 1$, so the channel transmission is determined by the length of the low-field section. For high drain bias, we conclude

$$
\mathcal{T}_{sat} = \frac{\lambda_{sat}}{\lambda_{sat} + \ell}.
\tag{7.21}
$$

We might expect that $\lambda_{lin} \approx \lambda_{sat}$ because in both cases, the important backscattering takes place in a low-field region.

Surprisingly, we find that $\mathcal{T}_{sat} > \mathcal{T}_{lin}$ because $\ell \ll L$. This is surprising because under high drain bias, carriers are more energetic in the

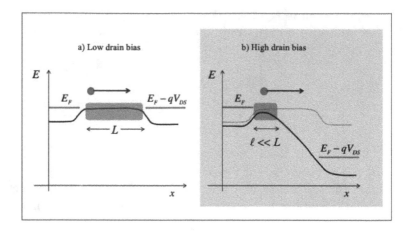

Fig. 7.5 Illustration of why the transmission depends on drain bias and why it is larger for high drain bias than for low drain bias. Left a): Low V_{DS}. Right b): High V_{DS}. The mean-free-path, λ, in the shaded regions is about the same in both cases.

high-field section, so they scatter more than under low drain bias. There is more scattering under high drain bias, but the transmission is higher because the transmission is determined by the low-field section, and under high drain bias, the length of the low-field section is much less than L. A MOSFET operates closer to the ballistic current limit under high drain bias than under low drain bias. It is also important to note that a MOSFET can operate close to the ballistic current limit even when there is a lot of scattering — if transport across the short bottleneck region is nearly ballistic. Scattering in the high-field region has little effect on the DC drain current.

7.5 Revisiting the virtual source model

Equation (4.25) introduced a simple VS model based on traditional MOSFET IV theory, and we mentioned that it produces excellent fits to the measured IV characteristics of nanoscale MOSFETs despite the fact that it is based on some dubious physics (e.g. the use of the scattering limited mobility, μ_n, and the high-field, scattering limited saturation velocity, υ_{sat}). Figure 4.9 was an example of the excellent fits that are obtained if μ_n and υ_{sat} are treated as fitting parameters. We now understand that these are not empirical fitting parameters but clearly-defined physical parameters.

By making the replacements, $\mu_n \rightarrow \mu_{app}$ and $v_{sat} \rightarrow v_{inj}$ and with the use of a better description of Q_n below and above threshold as discussed in Lecture 5, the VS model becomes a strongly physical, very accurate model for the IV characteristics of nanoscale MOSFETs. Our simple VS model for nanoscale MOSFETs is summarized as follows:

$$
\begin{aligned}
&I_D/W = |Q_n(0)| \langle v_x(0) \rangle \\
&Q_n(V_{GS}, V_{DS}) = -mC_g \frac{k_b T}{q} \ln\left(1 + e^{q(V_{GS}-V_T)/mk_BT}\right) \\
&V_T = V_{T0} - \delta V_{DS} \\
\\
&\langle v_x(0) \rangle = F_{SAT}(V_{DS})v_{inj} \\
&F_{SAT}(V_{DS}) = \frac{V_{DS}/V_{DSAT}}{\left[1 + (V_{DS}/V_{DSAT})^\beta\right]^{1/\beta}} \\
&V_{DSAT} = \frac{v_{inj}L}{\mu_{app}} \\
\\
&V_{GS} = V'_{GS} - I_D R_S \\
&V_{DS} = V'_{DS} - I_D(R_S + R_D)
\end{aligned}
\tag{7.22}
$$

Measured data is in terms of V'_{GS} and V'_{DS}, the voltages at the contacts. The model is in terms of voltages at the intrinsic device contacts, V_{GS} and V_{DS}, which are obtained by accounting for the series resistances. For $Q_n(V_{GS}, V_{DS})$, we use the empirical expression of Eq. (5.30), which describes the mobile charge from sub-threshold to above-threshold. This model is semi-empirical — most of the parameters in the model have a clear physical meaning, but β is an empirical parameter that is adjusted to match the drain saturation characteristics; it typically falls in a narrow narrow range of $\beta \approx 1.6 - 2.0$. Our VS model is very similar to the MIT VS model [4], which has a couple more empirical parameters to better fit measured data. The model can be used in two ways: 1) To compute accurate IV characteristics for well-designed nanoscale MOSFETs, or 2) to fit measured data and extract device-specific parameters that characterize the fabrication process.

The complete VS model gives IV characteristics that are continuous for V_{GS} below threshold to above and for V_{DS} from the linear to saturation region. Above threshold, the *IV* characteristics are described by

$$I_D(V_{GS}, V_{DS}) = WC_g (V_{GS} - V_T) \, v_{inj} \frac{V_{DS}/V_{DSAT}}{\left[1 + (V_{DS}/V_{DSAT})^\beta\right]^{1/\beta}} . \quad (7.23)$$

$$V_{DSAT} = v_{inj} L / \mu_{app}$$

With this simple model, we can compute surprisingly good MOSFET *IV* characteristics with only a few device model parameters.

To fit measured data, such as that shown in Fig. 4.9, we require some independently measured or specified parameters — C_g, L, and β. Six parameters are then adjusted to fit the measured data. The threshold voltage, V_{T0}, is adjusted to fit the measured off-current under low V_{DS}. The DIBL parameter, δ, is adjusted to fit the measured DIBL. (It also affects the output conductance.) The subthreshold slope parameter, m, is adjusted to fit the subthreshold slope (the MIT VS model also has a punchthrough parameter, m', to treat different subthreshold slopes under low and high V_{DS}). The apparent mobility, μ_{app}, is adjusted to fit the linear region slope of I_{DS} vs. V_{DS}, and the injection velocity, v_{inj}, is adjusted to fit the measured saturation currents. The series resistance, $R_{SD} = R_S + R_D$, affects both the linear and saturation regions and needs to be fitted too. Because the series resistance affects the linear and saturation regions differently, it is possible to independently deduce values for μ_{app} and R_{SD}. The result of the fitting process is a set of specific values for R_{SD}, V_{T0}, δ, m, μ_{app}, and v_{inj} for the device. it is also possible with careful analysis to deduce values for the ballistic injection velocity, v_T, the scattering limited mobility, μ_n, the mean-free-path, λ, the critical length, ℓ, as well as the transmission in the linear region, $\mathcal{T}_{lin}$, and in the saturation region, $\mathcal{T}_{sat}$. Reference [5] is an example of such an analysis.

7.6   Discussion

i) On the meaning of mobility at the nanoscale

The question of what mobility means in a nanoscale MOSFET needs some discussion. According to Eq. (6.16), the mobility is proportional to the mean-free-path. In transport theory, mobility is a material parameter that is well-defined under near-equilibrium conditions in a bulk material that is many mean-free-paths long (see Sec. 6.5 in [2] or Chapters 3 and 4 in [3]). It is not clear that mobility has any meaning at the nanoscale. In modern transistors, the channel length is comparable to the mfp, and under high drain bias, the carriers are very far from equilibrium. Nevertheless, device engineers find that the near-equilibrium mobility is strongly correlated to the performance of nanoscale transistors. How do we explain the relevance of mobility in nanoscale MOSFETs?

An equilibrium flux of carriers is injected into the channel from the source. Under low drain bias, these carriers remain near-equilibrium across the entire channel. For low V_{DS}, the transmission is controlled by the near-equilibrium mfp according to Eq. (7.5), but because μ_n is proportional to the near-equilibrium mfp, we can write $\mathcal{T}_{lin}$ in terms of the mobility as in Eq. (7.8). This explains the relevance of mobility for low V_{DS}.

Is mobility relevant for high V_{DS}? Under high drain bias, as the carriers gain energy in the drain field, their scattering rate increases, and the mfp decreases. As we have discussed, however, it is the low-field part of the channel that determines the overall channel transmission. Equation (7.21) shows how $\mathcal{T}_{sat}$ is related to the near-equilibrium mfp, λ. Since the mobility is also proportional to λ, $\mathcal{T}_{sat}$ can be written in terms of μ_n. Transport physics at the nanoscale is complex, but this argument captures the essence of the physics and explains the experimentally observed correlation of μ_n and I_{DSAT} in nanoscale MOSFETs.

We can get a quantitative estimate of the sensitivity of I_{DSAT} to mobility from the factor, $\mathcal{T}_{sat}/(2 - \mathcal{T}_{sat})$. Using Eq. (7.21) for $\mathcal{T}_{sat}$, we find

$$\frac{\mathcal{T}_{sat}}{(2 - \mathcal{T}_{sat})} = \frac{\lambda}{\lambda + 2\ell}. \tag{7.24}$$

According to Eq. (7.19), the injection velocity is

$$\begin{aligned} v_{inj} &= \left(\frac{\mathcal{T}_{sat}}{2 - \mathcal{T}_{sat}}\right) v_T = \frac{\lambda\, v_T}{\lambda + 2\ell} \\ &= \frac{1}{1/v_T + \ell/(\lambda v_T/2)}. \end{aligned} \tag{7.25}$$

Now, recall the definition of the diffusion coefficient, Eq. (7.4), $D_n = v_T \lambda/2$, which can be used to write the injection velocity as

$$v_{inj} = \left(\frac{1}{v_T} + \frac{1}{D_n/\ell}\right)^{-1}, \tag{7.26}$$

or

$$\boxed{\frac{1}{v_{inj}} = \frac{1}{v_T} + \frac{1}{D_n/\ell}.} \tag{7.27}$$

Equation (7.27) helps us understand carrier transport across the short bottleneck region of the channel that exists in the on-state (see Fig. 7.6b).
Carriers are injected into the virtual source, then they must diffuse across
the bottleneck region, but they cannot diffuse faster than the thermal velocity because diffusion is caused by random thermal motion. After diffusing
across the bottleneck, they encounter the high field portion of the channel,
which sweeps them across and out the drain. According to Eq. (7.27), the
injection velocity is less than the lower of the ballistic injection velocity and
D_n/ℓ, which is the velocity at which carriers diffuse across the bottleneck
region of length, ℓ. When ℓ is long or D_n small, $D_n/\ell \ll v_T$, and injection
velocity is the diffusion velocity. When ℓ is short or D_n large, $D_n/\ell \gg v_T$,
and the injection velocity is limited by the ballistic injection velocity. For
Si MOSFETs, the two terms on the RHS of Eq. (7.27) are roughly equal, so
a 10% increase in mobility (i.e. a 10% increase in D_n through the Einstein
relation) produces a 5% increase in I_{DSAT}.

ii) A few words about compact models

The term, *compact model*, is used in many fields of science and engineering
to describe a simple (usually analytical) model — as opposed to a simulation, which is typically a numerical solution to a set of equations. In
electronics, "compact model" typically refers to a model that describes an
electronic device in a form that is suitable for use by circuit designers in numerical circuit simulation programs. The compact model discussed in these
lectures has a different aim — to describe a device in terms of a few parameters with strong physical significance. These kinds of models are useful
for device characterization, process monitoring, for the conceptual understanding that guides device research, and for simple, hand calculations of
circuits.

BSIM (Berkeley Short Channel IGFET Model) is a widely-used compact
model available in all commercial circuit simulation programs [8]. The MIT
VS model and BSIM are much different types of compact models. The MIT
VS model for DC operation has 10 parameters — most with a very strong
physical meaning. Capacitances can be added to treat the AS small signal
operation. BSIM also describes the DC and AC small signal operation,
but it also describes transient operation and the noise characteristics as
well as a host of other factors as illustrated in Fig. 7.6. BSIM uses about
200 parameters to describe a nominal transistor as well as 700 additional

parameters to describe the transistor's "neighborhood" (e.g. the parasitic capacitances, the layout-dependent strain induced in the transistor, etc.) At its core, however, is a physical model analogous to the VS model.

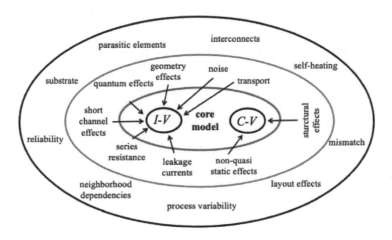

Fig. 7.6   IIustration of how a compact model for circuit design builds on core physical models for the *IV* and *CV* characteristics to model the comprehensive array of effects that need to be comprehended in modern circuit design. (After S. K. Saha [10].)

Developing compact circuit models requires a good understanding of the device physics, what goes on inside the circuit simulator, and the intended application. These types of compact models must satisfy the needs of circuit designers as well as the mathematical constraints of the circuit simulator. For an introduction to the science and art of developing compact models for circuit design, see [9] and [10].

7.7   Summary

In this lecture, we extended the ballistic treatment of the MOSFET discussed in Lecture 6 to include carrier scattering in the channel. The approach was based on the channel transmission coefficient,

$$
\begin{aligned}
\mathcal{T}_{lin} &= \frac{\lambda}{\lambda + L} &\quad (7.5)\\
\mathcal{T}_{sat} &= \frac{\lambda}{\lambda + \ell} &\quad (7.21)
\end{aligned}
$$
,

where λ is the near-equilibrium (low-field) mean-free-path for backscattering. Under low V_{DS}, the relevant length is the length of the channel, but under high V_{DS}, the relevant length is $\ell \ll L$, where ℓ is the length of the low-field, bottleneck section near the beginning of the channel.

The physics of how carrier scattering affect the IV characteristics is most clearly understood in terms of transmission, but we also showed that the results could also be expressed in the familiar traditional form. For I_{DLIN}, we found:

$$I_{DLIN} = \frac{W}{L}\mu_{app}|Q_n(V_{GS})|V_{DS} \tag{7.9}$$

$$\frac{1}{\mu_{app}} \equiv \frac{1}{\mu_n} + \frac{1}{\mu_B} \tag{7.10}$$

$$\mu_n \equiv \left(\frac{\upsilon_T \lambda}{2(k_B T/q)}\right) \tag{6.16}$$

$$\mu_B \equiv \left(\frac{\upsilon_T L}{2(k_B T/q)}\right) \tag{6.14}$$

$$\upsilon_T = \sqrt{\frac{2k_B T}{\pi m^*}} \tag{6.5}$$

I_{DLIN} is given by the traditional expression except that the mobility, μ_n is replaced by the apparent mobility, μ_{app}. The scattering limited mobility is proportional to the mfp, λ, and the ballistic mobility is proportional to the channel length, L. The apparent mobility controls I_{DLIN}, and it is less than the smaller of μ_n and μ_B. For a long channel MOSFET with $L \gg \lambda$, $\mu_B \gg \mu_n$, and $\mu_{app} \to \mu_n$. For a very short channel MOSFET with $L \ll \lambda$, $\mu_B \ll \mu_n$, and $\mu_{app} \to \mu_B$. Nanoscale Si MOSFETs operate in the quasi-ballistic regime where $\mu_B < \mu_{app} < \mu_n$.

We can also express the transmission result for I_{DSAT} in the traditional, velocity saturation form as:

$$I_{DSAT} = W|Q_n(V_{GS}, V_{DS})|\upsilon_{inj} \tag{7.19}$$

$$\upsilon_{inj} = \left(\frac{\mathcal{T}_{sat}}{2 - \mathcal{T}_{sat}}\right)\upsilon_T \leq \upsilon_T$$

I_{DSAT} is given by the traditional velocity saturation model with υ_{sat} replaced by the injection velocity, υ_{inj}. Carrier scattering reduces the injection velocity from its ballistic value, which is the thermal equilibrium unidirectional thermal velocity, υ_T. We also showed in this lecture that the VS model can be extended as to treat nanoscale MOSFETs. Equations (7.22) describe the IV characteristics for V_{DS} from the linear to saturation region, for V_{GS} from subthreshold to above threshold, and for channel lengths from the ballistic to quasi-ballistic regimes. Recall that for very long channel lengths, $I_{DSAT} \propto (V_{GS} - V_T)^2$; the very long channel length case is described by the classic square law MOSFET, not by the model developed in Lectures 6 and 7. In Lecture 6, we estimated the ballistic on-current for a Si MOSFET.

Under on-state conditions, $n_S \approx 10^{13} \, \text{cm}^{-2}$ and $v_T \approx 10^7 \, \text{cm/s}$, so for $W = 1 \, \mu\text{m}$, we found

$$I_{on} = W|Q_n|v_T = 10^{-4} \times (1.6 \times 10^{-19}) \times (10^{13}) \times 10^7 = 1.6 \quad \text{mA}.$$

This value is well above the 1 mA/μm on-current for a typical, n-channel Si MOSFET. From Eq. (7.19), we can deduce that $\mathcal{T}_{sat} \approx 0.75$ for a typical Si MOSFET, which is not far below the ballistic limit.

Our goal in these lectures has been to understand the physical operation of MOSFETs. The energy band approach discussed in Lecture 3 provides a very physical, qualitative explanation of a MOSFET's IV characteristics, and Lectures 6 and 7 translated this understanding into a simple, mathematical model that accurately describes modern MOSFETs with only a few parameters.

Lecture 7 Exercise: Virtual Source analysis of an n-channel MOSFET

The figure below shows measured IV characteristics of an Extremely Thin Silicon-On-Insulator (ETSOI) n-channel MOSFET with $L = 30$ nm [5]. The virtual source model fitting parameters are:

$$R_{SD0} = R_{S0} + R_{D0} = 130 \ \Omega - \mu m$$

$$\mu_{app} = 220 \ cm^2/V - s$$

$$\upsilon_{inj} = 0.82 \times 10^7 \ cm/s.$$

In addition, measurements of long channel transistors where $\mu_{app} = \mu_n$ give $\mu_n = 350 \ cm^2/V - s$.

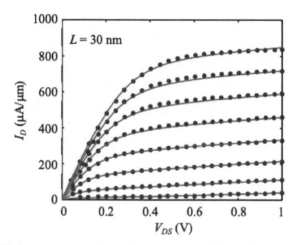

Measured IV characteristics of an $L = 30$ nm ETSOI MOSFET. The points are the measured data (similar to [5]), and the lines are the VS model fits. The first line is for $V_{GS} = -0.2$ V, and for each line above, V_{GS} increases by 0.1 V. The VS analysis and plot were provided by Dr. S. Rakheja, MIT, 2014. The data were provided by A. Majumdar, IBM, 2014. Used with permission.

From this data, make a "back of the envelope" estimate of the transmission in the linear and saturation regions.

The linear region transmission is given by Eq. (7.20), which can be expressed as

$$\mathcal{T}_{lin} = \frac{\lambda}{\lambda + L} = 1 - \frac{\mu_{app}}{\mu_n} = 0.37 \,.$$

For the saturation region transmission, we solve Eq. (7.19),

$$v_{inj} = \left(\frac{\mathcal{T}_{sat}}{2 - \mathcal{T}_{sat}}\right) v_T \leq v_T \, ,$$

for the transmission to find

$$\mathcal{T}_{sat} = \frac{2}{1 + v_T/v_{inj}} \, .$$

The injection velocity is known from fitting the VS model to the measured *IV* characteristics, but how can we determine the uni-directional thermal velocity from measured data?

From the definition of the apparent mobility, Eq. (7.10), we can solve for the ballistic mobility and find

$$\mu_B = \frac{\mu_{app}\,\mu_n}{\mu_n - \mu_{app}} = 592 \ \mathrm{cm}^2/\mathrm{V-s} \, ,$$

but we also know that the ballistic mobility is related to v_T by Eq. (6.14), so we find

$$\mu_B \equiv \left(\frac{v_T L}{2(k_B T/q)}\right) \rightarrow v_T = \frac{2(k_b T/q)\mu_B}{L} = 1.03 \times 10^7 \ \mathrm{cm/s} \, .$$

Finally, we estimate the transmission in saturation as

$$\mathcal{T}_{sat} = \frac{2}{1 + v_T/v_{inj}} = \frac{2}{1 + 1.03/0.82} = 0.89 \, .$$

As expected (for reasons summarized in Fig. 7.5), we find $\mathcal{T}_{lin} < \mathcal{T}_{sat}$. The MOSFET operates closer to the ballistic current limit under high V_{DS} where there is more scattering in the device than under low V_{DS}.

Finally, we can estimate ℓ, the length of the bottleneck region. From Eq. (6.16) for μ_n, we can estimate the mean-free-path from the measured mobility as

$$\lambda = \frac{2(k_B T/q)\mu_n}{v_T} \approx 18 \ \mathrm{nm} \, .$$

From Eq. (7.21) for $\mathcal{T}_{sat}$, we can estimate ℓ as

$$\ell = \lambda\,(1/\mathcal{T}_{sat} - 1) \approx 2 \ \mathrm{nm} \, ,$$

so, ℓ is roughly 10% of the channel length. The bottleneck for current at the beginning of the channel is very short.

The analysis in Sec. 19.5 of [1] produces similar, but not identical results. The difference is that the analysis here is only in terms of parameters deduced from the measured *IV* characteristics. Both analyses assume non-degenerate carrier statistics and other details, so the results should be considered to be estimates that give a feel for the typical numbers to be expected.

7.8 References

For a more comprehensive discussion of the transmission theory of the MOSFET including the use of Fermi-Dirac statistics and the analysis of measured IV characteristics, see:

[1] Mark Lundstrom, *Fundamentals of Nanotransistors*, World Scientific Publishing Co., Singapore, 2018.

For more on the physics of carrier scattering, the mean-free-path for backscattering, and transmission, see Lecture 6 in

[2] Mark Lundstrom and Changwook Jeong, *Near-Equilibrium Transport: Fundamentals and Applications*, World Scientific Publishing Company, Singapore, 2012.

For more about the physics of carrier scattering and Matthiessen's Rule for combining mobilities, see:

[3] Mark Lundstrom, *Fundamentals of Carrier Transport*, 2nd Ed., Cambridge Univ. Press, Cambridge, U.K., 2000.

The MIT Virtual Source Model is described in the first reference below, and the second reference below is an example of its use in analyzing measured IV data. The final two references describe extensions to the original VS model.

[4] A. Khakifirooz, O.M. Nayfeh, and D.A. Antoniadis, "A Simple Semiempirical Short-Channel MOSFET Current-Voltage Model Continuous Across All Regions of Operation and Employing Only Physical Parameters," *IEEE Trans. Electron. Dev.*, **56**, pp. 1674-1680, 2009.

[5] A. Majumdar and D.A. Antoniadis, "Analysis of Carrier Transport in Short-Channel MOSFETs," *IEEE Trans. Electron. Dev.*, **61**, pp. 351-358, 2014.

[6] Shaloo Rakheja, Mark Lundstrom, and Dimitri Antoniadis, "An Improved Virtual-Source-Based Transport Model for Quasi-Ballistic Transistors - Part I: Capturing Effects of Carrier Degeneracy, Drain-Bias Dependence of Gate Capacitance, and Non-linear Channel-Access Resistance," *IEEE Trans. Electron Dev.*, **62**, pp. 2786-2793, 2015.

[7] Shaloo Rakheja, Mark Lundstrom, and Dimitri Antoniadis, "An Improved Virtual-Source-Based Transport Model for Quasi-Ballistic Transistors - Part II: Experimental Verification," *IEEE Trans. Electron Dev.*, **62**, pp. 2794-2801, 2015.

Integrated circuit chip designers use very comprehensive, highly accurate compact models. BSIM in an industry-standard compact model. Tsividis and McAndrew provide a good introduction to these kinds of compact models and Saha discusses several different models as well as practical considerations for efficient design.

[8] "BSIM," https://en.wikipedia.org/wiki/BSIM, July 4, 2021.

[9] Y. Tsividis and C. McAndrew, *Operation and Modeling of the MOS Transistor*, 3rd Ed., Oxford Univ. Press, New York, 2011.

[10] S.K. Saha, *Compact Models for Integrated Circuit Design: Conventional Transistors and Beyond*, CRC Press, Taylor and Francis Group, Boca Raton, Florida, 2016.

Lecture 8

Bulk MOSFETs

8.1 Introduction
8.2 PN junctions
8.3 MOS band bending in pn junctions
8.4 MOS electrostatics normal to the channel
8.5 Surface potential at the onset of inversion
8.6 Threshold voltage and body effect
8.7 Mobile charge vs. gate voltage
8.8 Summary
8.9 References

8.1 Introduction

Figure 1.2 showed a scanning electron micrograph cross section of a bulk (or "planar") Si MOSFET circa 2000. As shown in Fig. 8.1a, n-channel bulk MOSFETs are built on p-type Si, but as shown in Fig. 3.11, leading-edge MOSFETs for digital electronics now have much different, non-planar structures. Non-planar MOSFETs have very thin, nominally undoped Si channels with the gate wrapped around much, if not all, of the channel to control short channel effects. In these kinds of *fully-depleted* or *ultra-thin body* MOSFETs (Fig. 8.1b) the entire region between the source and drain is depleted of majority carriers. Our discussion of the mobile charge vs. gate voltage in Lecture 5 assumed a fully-depleted, ultra-thin body structure because that best represents modern, leading edge MOSFETs for digital electronics, and it also simplifies the mathematical analysis. Bulk MOSFETs are still, however, widely-used. This lecture discusses how their *IV* characteristics differ from fully-depleted MOSFETs.

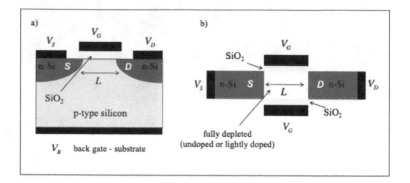

Fig. 8.1   Idealized sketches of: a) a bulk (or planar) n-channel MOSFET and b) a fully-depleted, ultra-thin body MOSFET. The fully-depleted structure is an idealization of a FinFET in which the channel is a vertical "fin" of Si, and the gate is wrapped around three sides of the channel (see Fig. 3.11).

The IV characteristics of bulk and fully depleted MOSFETs are very similar, but there are two differences to be aware of. First, note from Fig. 8.1 that the bulk MOSFET is a four terminal device. The body of the MOSFET (the p-type bulk) is at a potential, V_B. The body contact is sometimes referred to as a *back gate*. The voltage between the source and the body, $V_{SB} = V_S - V_B$ affects the threshold voltage of the MOSFET. The source and the body might both be grounded, in which case the body has no effect, but in circuits that stack MOSFETs in series, the source and body may be at different voltages, and the dependence of V_T on V_{SB} can have important effects. Bulk MOSFETs are still used in many applications, so it is important to understand $V_T(V_{SB})$, which is called the *body effect*.

The second difference between bulk and fully depleted MOSFETs is a small, but important difference in the Q_n vs. V_{GS} characteristic in the subthreshold region. Both effects will be discussed in this lecture.

8.2   PN junctions

The source-to-body junction is a pn junction, so we should first review some fundamentals of pn junctions. Figure 8.2 shows the equilibrium energy band diagram. Think of the n-type region as the source and the p-type region as the part of the channel near the source. On the left are the individual semiconductors. As shown on the right, when we conceptually bring the two semiconductors together, electrons transfer from the n-type semiconductor with the higher E_F to the p-type semiconductor with the lower E_F, and holes transfer from the p-side to the n-side. The charge transfer increases the electrostatic potential of the n-type semiconductor. (Recall that a positive electrostatic potential lowers an electron's energy.) After the charge transfer process is complete, electrons see an energy barrier, qV_{bi}, that prevents further diffusion to the p-side, and holes see an energy barrier that keeps them on the p-type side (hole energy increases downward on this electron energy plot). The electrostatic potential on the n-side of the pn junction is V_{bi} volts higher than on the p-side, which lowers the bands on the n-side so that the two Fermi levels align, and there is one, spatially uniform Fermi level in equilibrium. The *built-in potential* difference between the two sides of the junction is

$$qV_{bi} = E_{Fn} - E_{Fp}. \tag{8.1}$$

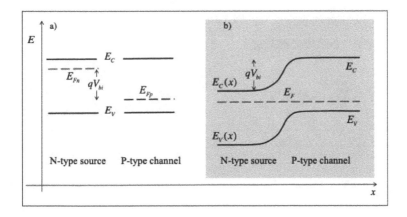

Fig. 8.2 Equilibrium energy band diagrams for pn junctions. a) Separate n-type and p-type semiconductors. b) The resulting equilibrium energy band diagrams when the two semiconductors are conceptually joined.

In a MOSFET, there is a gate above the p-region. A positive voltage on the gate increases the electrostatic potential, ψ, in the p-type Si under the gate. The next step is to understand how a gate voltage affects the source to channel pn junction.

8.3   MOS band bending in PN junctions

Figure 8.3 shows energy band diagrams for the source to channel pn junction. The plots are energy vs. position, x, along the channel at $y = 0$, the surface of the silicon. Only the p-region near the source is shown, and a bias of $V_S = V_B = 0$ is assumed. A positive voltage on the gate, which is located above the p-side, raises the potential in the channel (the p-region), and, therefore, lowers the energy bands of the p-region. Energy band diagrams for two different gate voltages are shown.

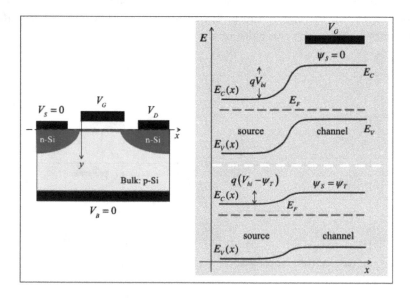

Fig. 8.3   Energy band diagrams along the channel from the n-type source to the p-type channel. Upper right: Flat band conditions for which $V_{GS} = V_{FB}$, $\psi_s = 0$, and there is no band bending normal to the channel. Lower right: $V_G = V_T$, $\psi_s = \psi_T$. The drain pn junction is not shown, and for this case, $V_S = V_B = 0$ V.

Consider first the figure on the upper right. The electrostatic potential at the surface ($y = 0$) of the p-region is ψ_s, the *surface potential*. Deep in the bulk of the p-type Si, the electrostatic potential is $\psi = 0$, so for this case, there is no band bending in the y-direction. This gate voltage (below threshold) produces *flat band* conditions.

The lower right figure shows what happens for a gate voltage of $V_G = V_T$. As we'll discuss in Sec. 8.5, at the onset of conduction, the surface potential is $\psi_s = \psi_T > 0$. The positive gate voltage has produced a positive surface potential of ψ_T, which has lowered the energy barrier between the source and channel to $q(V_{bi} - \psi_T)$, so that current can flow.

Consider next what happens when we reverse bias the source-to-body junction. As shown in Fig. 8.4a for the case where $\psi_s = 0$ in the p-type Si, a positive voltage on the source, $V_S > 0$ with $V_B = 0$, lowers the electron quasi-Fermi level and the bands in the source and increases the source to channel energy barrier from qV_{bi} to $q(V_{bi} + V_S)$. As shown in Fig. 8.4b, a much larger gate voltage is needed to increase the surface potential to $\psi_T + V_S$ so that the energy barrier has the same small value needed to turn the transistor on when $V_S = 0$.

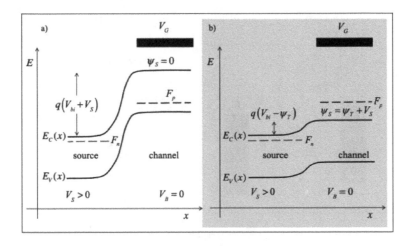

Fig. 8.4 Energy band diagrams along the channel from the n-type source to the p-type channel. In this case, $V_S > 0$ and $V_B = 0$ V, so the source to body pn junction is reverse biased. a) Flat band conditions for which $V_G = V_{FB}$, $\psi_s = 0$, and there is no band bending normal to the channel. b) $V_{GS} = V_T$, $\psi_s = \psi_T + V_S$.

Let's summarize. To turn on an n-channel MOSFET, we require

$$V_{GS} = V_G - V_S > V_T. \tag{8.2}$$

If $V_S = 0$, we need a certain gate voltage to make $\psi_s = \psi_T$ in the p-type semiconductor. If $V_S > 0$, we need to increase ψ_s in the p-type Si from ψ_T to $\psi_T + V_S$, and to do so requires a larger gate voltage. If the new gate voltage was simply the gate voltage with $V_S = 0$ plus V_S, then V_{GS} would not change, so V_T would not change. It turns out, however, that V_G must increase by more than V_S to increase ψ_s from ψ_T to $\psi_T + V_S$. To understand why, we need to understand the electrostatics in the direction normal to the channel.

8.4   MOS electrostatics normal to the junction

We have been drawing energy band diagrams along the direction of the channel, the x-direction in Fig. 8.4. Now let's draw energy band diagrams in the y-direction, normal to the channel. Figure 8.5 shows equilibrium energy band diagrams for three different gate voltages. A p-type semiconductor, which is relevant for an n-channel MOSFET, is assumed. The gate oxide blocks current flow, so the metal and the semiconductor are separately in equilibrium with their own Fermi levels.

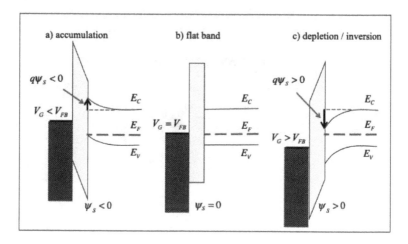

Fig. 8.5   Energy band diagrams normal to the channel from the gate to the bulk, p-type Si for three different gate voltages.

Shown in Fig. 8.5a is the case for which the gate voltage is less than the flat band voltage, $V_G < V_{FB}$. (In an ideal structure, $V_{FB} = 0$.) When $V_G < V_{FB}$, the surface potential, ψ_s, is less than zero, the energy bands are pulled up, the valence band is closer to the Fermi level at the surface, so holes pile up, or *accumulate*, at the surface. The conduction band is further from the Fermi level at the surface, so the electron concentration at the surface is less than in the bulk. Alternatively, we can think of the negative voltage on the gate as putting a negative charge on the gate, which attracts holes and pushes electrons away from the surface.

Shown in Fig. 8.5b is the flat band case for which $V_G = V_{FB}$ and $\psi_s = 0$. Finally, shown in Fig. 8.5c is the case for which $V_G > V_{FB}$ and $\psi_s > 0$. The energy bands are pulled down, the valence band is further from the Fermi level at the surface, so the hole concentration at the surface is less than in the bulk. At the same time, the conduction band is closer to the Fermi level at the surface, so electrons pile up at the surface. We can think of the positive charge on the gate as pushing holes away and attracting electrons to the surface.

For n-channel MOSFETs, the positive surface potential region is the region of interest. Initially, holes are pushed away, and we say that the semiconductor is *depleted* of majority carrier holes. When holes are pushed away from the surface, they leave behind negatively-charged, ionized acceptors. The electron density increases exponentially with increasing ψ_s. Until ψ_s reaches a critical value, ψ_T, however, the electron density is small compared to the doping density. This range, of surface potentials, $0 < \psi_s < \psi_T$, is called the *depletion region*. The space charge density in the depleted region near the surface is $\rho \approx -qN_A^-$ C/cm^3 because the electron concentration can be ignored. When $\psi_s > \psi_T$, the electron concentration at the surface is higher than the hole concentration in the bulk, and we say that the semiconductor is *inverted*.

Consider the depletion region, $0 < \psi_s < \psi_T$, where the electron concentration is too small to have a significant effect on the electrostatics. We encountered the integral form of Gauss's Law in Eq. (5.10); the differential form in 1D is

$$\frac{dD(y)}{dy} = \frac{d(\kappa_s \epsilon_0 \mathcal{E})}{dy} = \rho(y) \,, \tag{8.3}$$

where D is the displacement field, $\mathcal{E}$ the electric field, and ρ is the space charge density. In the depletion region, the space charge density is dominated by ionized acceptors. Assuming complete ionization, $\rho \approx -qN_A$, so

$$\frac{d\mathcal{E}}{dy} = -\frac{qN_A}{\kappa_s \epsilon_0} \,, \tag{8.4}$$

which tells us that in depletion the electric field vs. position is linear with a negative slope. Assuming that majority carrier holes are pushed away from the surface to a depth, W_D, the depletion layer thickness, we can sketch the electric field vs. position in depletion as shown in Fig. 8.6.

From Fig. 8.6 and Eq. (8.4), we see that

$$\frac{\mathcal{E}_S}{W_D} = \frac{qN_A}{\kappa_s \epsilon_0} \,. \tag{8.5}$$

The electrostatic potential is the integral of the electric field vs. position. The potential drop across the semiconductor, the surface potential, ψ_s, is the area under the curve in Fig. 8.6. The result is

$$\psi_s = \frac{1}{2}\mathcal{E}_S W_D \,. \tag{8.6}$$

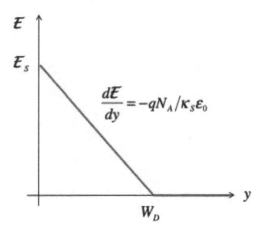

Fig. 8.6   Electric field vs. position for a p-type semiconductor in depletion.

For a given surface potential, Eqs. (8.5) and (8.6) are two equations for the two unknowns, $\mathcal{E}_S$ and W_D. Solving for W_D, we find

$$W_D = \sqrt{\frac{2\kappa_s\epsilon_0\psi_s}{qN_A}}.$$
(8.7)

Our goal in doing these calculations is to find the total, areal charge in the semiconductor,

$$Q_S = \int_0^\infty \rho(y)dy = Q_D + Q_n \quad \text{C/m}^2\,,$$
(8.8)

where Q_D is the the depletion charge per m^2 due to ionized acceptors, and Q_n is the charge per m^2 due to mobile electrons. In depletion, the charge lies between $x = 0$ and $x = W_D$ and consists only of ionized acceptors because the mobile charge density is negligible. We find

$$Q_D = -qN_AW_D = -\sqrt{2qN_A\kappa_s\epsilon_0\psi_s} \quad \text{C/m}^2\,.$$
(8.9)

8.5   Surface potential at the onset of inversion

Before we can compute V_T, we need to define the surface potential, ψ_T, at the onset of inversion. It should be understood that there is no sharp transition between off and on in a MOSFET. When Q_n becomes large enough,

we define the MOSFET to be "on". In Sec. 5.4 for the fully depleted MOS-
FET, we argued what when the potential in the semiconductor is large
enough to bring the conduction band to the Fermi level, the MOSFET
turns on (Eq. (5.20)). For a bulk MOSFET, however, the transition between off and on has been traditionally defined in terms of the doping of
the p-type bulk.

Figure 8.7 shows an energy band diagram for a bulk MOSFET at the
onset of conduction. Deep in the bulk, the p-type semiconductor is neutral
with $p_0(x \to \infty) = p_B = N_A$; the Fermi level lies below the intrinsic level
by an amount $q\psi_B$. The hole concentration in the bulk is related to the
Fermi level by

$$p_B = n_i e^{(E_i - E_F)/k_B T} = N_A , \qquad (8.10)$$

which can be solved for

$$(E_i - E_F)_{bulk} \equiv q\psi_B = k_B T \ln (N_A/n_i) . \qquad (8.11)$$

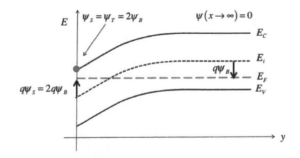

Fig. 8.7 Band-bending in an n-channel MOSFET at the onset of conduction. The
direction, y is normal to the channel, and the Si is p-type. A positive surface potential
sufficient to invert the surface has been applied.

Figure 8.7 shows what happens when $\psi_s = 2\psi_B$. At the surface, the
Fermi level is above the intrinsic level by the same amount that it was below
the intrinsic level in the bulk. This means that the electron concentration
at the surface is the same as the hole concentration in the bulk. We say
that the semiconductor is *inverted* — the bulk is p-type and the surface is
n-type. We take this surface potential as the boundary between the off and
on states of the bulk MOSFET,

$$\psi_s = \psi_T = 2\psi_B , \qquad (8.12)$$

which is analogous to Eq. (5.20) for the fully-depleted MOSFET.

8.6   Threshold voltage and body effect

To compute the threshold voltage, we begin with

$$V_G = -\frac{Q_S(\psi_s)}{C_{ox}} + \psi_s \,, \tag{8.13}$$

which is analogous to Eq. (5.14) for the fully depleted MOSFET except that we use the total charge in the semiconductor, Q_S. (For the fully depleted MOSFET, there is no depletion charge so $Q_S = Q_n$.) The bulk MOSFET turns on when $\psi_s = \psi_T$; at this point, we assume that the mobile charge is still negligible, so for a bulk MOSFET at $\psi_s = \psi_T$, $Q_S = Q_D + Q_n \approx Q_D$. We conclude that for a bulk MOSFET, the gate voltage at the onset of conduction is

$$V_G = V_{T0} = -\frac{Q_D(\psi_T)}{C_{ox}} + \psi_T \,. \tag{8.14}$$

We call this threshold voltage, V_{T0} because we are assuming that $V_S = V_B = 0$. Equation (8.14) is analogous to Eq. (5.21) for the fully-depleted MOSFET, but for the fully-depleted MOSFET, the channel is undoped and $Q_D = 0$.

Assume next that $V_S > 0$ and to keep the bookkeeping simple, assume that $V_B = 0$. Figure 8.4 showed that for this case, the surface potential must be $\psi_s = \psi_T + V_S$. Equation (8.14) becomes

$$V_G = V_T = -\frac{Q_D(\psi_T + V_S)}{C_{ox}} + \psi_T + V_S \,. \tag{8.15}$$

Note that the gate voltage at the onset of inversion is greater than Eq. (8.14) plus V_S because $Q_D(\psi_T + V_S) > Q_D(\psi_S)$. The difference is what increases V_T.

The threshold voltage is the gate to source voltage needed to turn the MOSFET on, so from Eq. (8.15)

$$V_T = V_G - V_S = \frac{Q_D(\psi_T + V_S)}{C_{ox}} + \psi_T \,, \tag{8.16}$$

which can be written as

$$V_T = V_{T0} + \frac{[Q_D(\psi_T + V_S) - Q_D(\psi_T)]}{C_{ox}} \,. \tag{8.17}$$

The voltage on the body contact may not always be at $V_B = 0$, but it is only the source to body voltage that matters, so V_S in Eq. (8.17) should in general be replaced by V_{SB}.

The depletion charge is given by Eq. (8.9) and the surface potential at the onset of inversion is $\psi_T = 2\psi_B$ from Eq. (8.12), so with these equations and Eq. (8.17), we can evaluate V_T, but first these is an adjustment that must be made.

We have been assuming an "ideal" gate electrode for which the Fermi level in the metal gate lines up with the Fermi level in the Si when $V_G = 0$. We saw for the pn junction that whenever we bring two conductors with different Fermi levels together a *contact potential* or built-in potential results. For MOS structures, the result is that even for $V_G = 0$, there is a built-in potential related to the difference in the Fermi levels of the isolated metal and semiconductor (see Eq. (8.1)). A *flat-band voltage*, V_{FB} must be applied to the gate to produce flat band conditions in the semiconductor. (See Sec. 7.1 in [1] for a discussion of how metal-semiconductor workfunction differences produce a non-zero flatband voltage.) The result is that we need to add this flat band correction to Eq. (8.17).

Putting these results together, Eq. (8.17) gives the threshold voltage of a bulk MOSFET as

$$V_T = V_{T0} + \gamma \left[\sqrt{2\psi_B + V_{SB}} - \sqrt{2\psi_B} \right]$$
$$V_{T0} = V_{FB} + \frac{\sqrt{2qN_A\kappa_s\epsilon_0(2\psi_B)}}{C_{ox}} + 2\psi_B \qquad (8.18)$$
$$\gamma = \frac{\sqrt{2qN_A\kappa_s\epsilon_0}}{C_{ox}},$$

where γ is the *body effect parameter*. The dependence of the threshold voltage on the source to body voltage can have significant effects on circuit performance.

8.7 Mobile charge vs. gate voltage

The body effect is the main difference between bulk and ultra thin body, fully depleted MOSFETs, but there is one other difference that we should briefly discuss. For both bulk and fully depleted MOSFETs, we find that above threshold

$$Q_n = -C_g \left(V_{GS} - V_T \right), \qquad (8.19)$$

and below threshold,

$$Q_n \propto \exp\left[q(V_{GS} - V_T)/mk_BT \right]. \qquad (8.20)$$

The parameter, $m \geq 1$, is important because as discussed in Sec. 3.7, the subthreshold swing, which should be as small as possible, is proportional to m (see Eq. (3.14)).

As discussed in Lecture 3, the parameter m is increased by short channel effects. An m greater than one means that the gate does not have full control of the potential at the beginning of the channel. Recall that the drain voltage can also lower the energy barrier. This drain-induced barrier lowering (DIBL) increases m. In principle, however, for fully depleted MOSFETs if we can suppress short channel effects with non-planar structures such as those shown in Fig. 3.11, then we can achieve $m = 1$. This is not the case for bulk MOSFETs for which $m > 1$ even when short channel effects are fully suppressed. To understand why, we must look closely at the gate voltage - surface potential relation in the subthreshold where the semiconductor is in depletion.

The relation between gate voltage and surface potential is given by Eq. (8.13), but this equation doesn't give a simple relation between the surface potential and the gate voltage. We can develop a simple, approximate relation between ψ_s and V_G from the two capacitor model shown in Fig. 8.8. The first capacitor is the capacitance of the oxide layer, and the second is the capacitance of the semiconductor. In depletion, the semiconductor capacitance is the depletion capacitance, which is readily computed from the surface potential. Note that the depletion capacitance is a nonlinear function of ψ_s, so to keep the analysis simple, we assume an average C_D computed with an average ψ_s in depletion.

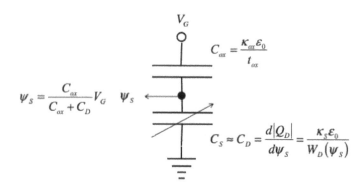

Fig. 8.8   A simple, two capacitor approximate solution to Eq. (8.13).

Elementary circuit analysis, voltage division for two capacitors in series, gives ψ_s in terms of V_G. The result can be written as

$$\psi_s = V_G/m\,,\qquad(8.21)$$

where

$$m = 1 + C_D/C_{ox} > 1\,.\qquad(8.22)$$

Equation (8.21) is analogous to Eq. (3.12), but note that the analysis here is one-dimensional and that for a bulk MOSFET, $m > 1$ even in the absence of 2D electrostatics. The fact that for a fully depleted, ultra-thin body MOSFET, m can be closer to one than for a bulk MOSFET is an important advantage because it is essential to keep the subthreshold swing as small as possible to permit low voltage (i.e. low power) operation.

8.8   Summary

The *IV* characteristics of bulk (planar) and fully-depleted ultra-thin body (nonplanar) MOSFETs are similar, but bulk MOSFETs have a fourth terminal, the body contact that gives rise to the body effect, which has significant effects on circuits. The effect of source-to-body voltage on the threshold voltage of a bulk MOSFET is described by the following equations.

$$
\begin{array}{l}
V_T = V_{T0} + \gamma \left[\sqrt{2\psi_B + V_{SB}} - \sqrt{2\psi_B} \right] \qquad (8.18) \\[2mm]
V_{T0} = V_{FB} + \dfrac{\sqrt{2qN_A\kappa_s\epsilon_0(2\psi_B)}}{C_{ox}} + 2\psi_B \\[2mm]
\gamma = \dfrac{\sqrt{2qN_A\kappa_s\epsilon_0}}{C_{ox}} \\[2mm]
\psi_B = (k_BT/q)\ln(N_A/n_i) \qquad (8.11)
\end{array}
$$

The fact that fully-depleted, ultra-thin body MOSFETs are capable of lower subthreshold swing is an important advantage for large scale digital electronics. The relevant equations for bulk MOSFETs are

$$
\begin{array}{l}
m = 1 + C_D/C_{ox} > 1 \qquad (8.22) \\[2mm]
C_D = \dfrac{\kappa_s\epsilon_0}{W_D} \quad \text{F/cm}^2 \\[2mm]
C_{ox} = \dfrac{\kappa_{ox}\epsilon_0}{t_{ox}} \quad \text{F/cm}^2 \\[2mm]
SS = 2.3m(k_BT/q) \ \text{mV/decade} \qquad (3.14)
\end{array}
$$

Finally, Fig. 8.9 shows the schematic circuit symbols for bulk MOS-FETs. There is no universal convention for drawing these symbols, but this figure shows a commonly used approach. On the top are the symbols for n- and p-channel MOSFETs that we used in Lecture 2. The dashed lines for the channels indicate that these are enhancement mode (normally off) devices — a channel does not exist until the correct V_{GS} is applied. As shown on the bottom, bulk MOSFETs are four terminal devices; the fourth terminal is the body, or bulk contact. The direction of the arrow indicates the type of device. For an n-channel MOSFET, the arrow points from the p-type bulk to the n-type channel while for a p-channel MOSFET, the arrow points from the p-type channel to the n-type bulk.

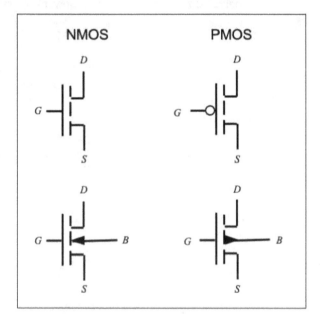

Fig. 8.9   Comparison of the circuit schematic symbols for three-terminal n- and p-channel MOSFETs (top) and for four-terminal, bulk MOSFETS (bottom).

Lecture 8 Exercise: Typical Numbers for Bulk MOSFETs

This exercise will help you get calibrated on some typical numbers for a bulk Si MOSFET at 300 K. Assume the following:

$$t_{ox} = 1.2 \text{ nm}$$
$$N_A = 10^{18} \text{ cm}^{-3}$$
$$n_i = 10^{10} \text{ cm}^{-3}$$
$$V_{FB} = -1.0 \text{ V}$$
$$\kappa_{ox} = 4$$
$$\kappa_s = 12$$
$$\epsilon_0 = 8.854 \times 10^{-14} \text{ F/cm}$$
$$k_B T/q = 0.026 \text{V}$$

1) Compute the surface potential at the onset of conduction

From Eq. (8.12):

$$\psi_T = 2\psi_B .$$

From Eq. (8.11):

$$\psi_B = (k_B T/q) \ln(N_A/n_i) = 0.026 \ln(10^8) = 0.48 \text{ V},$$

so we find

$$\psi_T = 2\psi_B = 0.96 \text{ V}.$$

2) Compute the width of the depletion layer at the onset of conduction

From Eq. (8.7):

$$W_D = \sqrt{\frac{2\kappa_s \epsilon_0 \psi_T}{q N_A}} = \sqrt{\frac{2 \times 12 \times 8.854 \times 10^{-14} \times (0.96)}{1.6 \times 10^{-19} \times 10^{18}}} = 3.6 \times 10^{-6} \text{ cm}.$$

3) Compute the threshold voltage when the source and body are at the same voltage.

From Eq. (8.14):

$$V_{T0} = V_{FB} + \frac{qN_AW_D}{C_{ox}} + \psi_T.$$

$$C_{ox} = \frac{\kappa_{ox}\epsilon_0}{t_{ox}} = \frac{4 \times 8.854 \times 10^{-14}}{1.2 \times 10^{-7}} = 3.0 \times 10^{-6} \ \text{F/cm}^2.$$

$$V_{T0} = -1.0 + \frac{1.6 \times 10^{-19} \times 10^{18} \times 3.6 \times 10^{-6}}{3.0 \times 10^{-6}} + 0.96 = 0.15 \ \text{V}.$$

4) Compute the threshold voltage when the source to body voltage is 0.6 V.

From Eq. (8.18):

$$V_T = V_{T0} + \gamma\left[\sqrt{\psi_T + V_{SB}} - \sqrt{\psi_T}\right].$$

$$\gamma = \frac{\sqrt{2qN_A\kappa_s\epsilon_0}}{C_{ox}} = \frac{5.8 \times 10^{-7}}{3 \times 10^{-6}} = 0.19 \ \sqrt{\text{V}}.$$

$$V_T = 0.15 + 0.19\left[\sqrt{0.96 + 0.6} - \sqrt{0.96}\right] = 0.15 + 0.05 = 0.20 \ \text{V}.$$

Because of the body effect, the threshold voltage has increased from 0.15 V to 0.20 V.

5) Compute the subthreshold swing.

From Eq. (3.14):

$$SS = 2.3m(k_BT/q) = m \times 60 \ \text{mV/decade}.$$

From Eq. (8.22):

$$m = 1 + C_D/C_{ox} = 1 + (\kappa_s\epsilon_0/W_D)/C_{ox} = 1 + 0.1,$$

so we find

$$SS = 1.1 \times 60 = 66 \ \text{mV/decade}.$$

Note that $m > 1$ even though we have not included the effects of 2D electrostatics, which would increase m further.

8.9   References

For a more thorough discussion of the depletion approximation, see Lecture 6 in the following text. For more about the mobile charge in a bulk MOS-FET, see Lecture 8.

[1] Mark Lundstrom, *Fundamentals of Nanotransistors*, World Scientific Publishing Co., Singapore, 2018.

For an excellent discussion of the body effect including some nice 2D energy band diagrams, see Chapter 10 in the classic book by Andy Grove.

[2] A.S. Grove, *Physics and Technology of Semiconductor Devices*, John Wiley and Sons, Inc., New York, 1967.

Lecture 9

Power MOSFETs

9.1 Introduction

Power MOSFETs are the most common power semiconductor devices, but to withstand high voltages, high currents, and high power dissipation, special designs are required. The LDMOS (laterally-diffused metal-oxide semiconductor) transistor is a power MOSFET that can be implemented in standard CMOS digital processes to produce *mixed signal* integrated circuits (ICs) in which information processing is done digitally with CMOS and analog functions such as wireless communication with RF power MOSFETs on the same chip. Power MOSFETs are also used as switches for power conversion and transmission and to control motors, relays, and lights [1, 2]. Some power electronic systems can benefit by having the digital control circuitry and power devices on the same chip, but discrete power devices are widely used when the applications demand high voltages and currents.

For electric power control and conversion, the MOSFET is used as a switch. As shown in Fig. 9.1, when the transistor is on, the switch should provide a high current, I_{on}, at a low voltage, V_{on}, which means that a low *on-resistance*, R_{on}, is needed. When the transistor is off, the switch should be able to withstand a high voltage, V_{br}, the *blocking voltage*, before it breaks down and conducts. For RF power applications high transconductance, high output resistance, and good high frequency performance are needed along with high current and voltage capabilities.

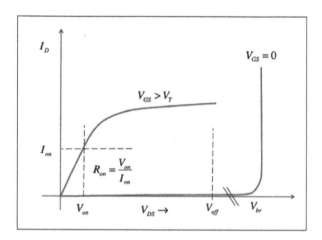

Fig. 9.1   IV characteristics of a power MOSFET illustrating the on-resistance, R_{on}, and drain breakdown voltage, V_{br}, which is called the blocking voltage. When the switch is closed, $V_{DS} = V_{on}$ and $I_D = I_{on}$. When the switch is off, $V_{DS} = V_{off}$ and $I_D \approx 0$.

Power MOSFETs can be implemented with lateral structures in which the current flows along the surface of the wafer or vertical structures in which the current flows from the top to the bottom of the wafer. Lateral structures like LDMOS are well-suited for mixed-signal, analog/digital ICs. They are used in audio power amplifiers and in RF power amplifiers for cellular communication. The DMOSFET and the "trench" UMOSFET are discrete, vertical devices used as switches in power electronics [1, 2]. Other types of power semiconductor devices include bipolar junction transistors (BJTs) (to be discussed in Lecture 11, Sec. 5) and insulated gate bipolar junction transistors (IGBTs). For comprehensive treatments of power semiconductor devices, see [1, 2].

In this lecture, we discuss two power MOSFET designs that illustrate the fundamental trade-off between on-resistance and breakdown voltage and the concept of a safe operating area (SOA), which are relevant to all power semiconductor devices.

9.2 Device structures

Figure 9.2 is a cross-section of a simple, n-channel LDMOS transistor that can be integrated into a CMOS process [3]. Many variations on the basic design exist with the device architecture and doping profiles selected to achieve application-specific targets for integration density, power-handling capabilities, and reliability. A device of this kind may have a blocking voltage on the order of 50 V. Unlike the MOSFETs we have discussed previously, the device in Fig. 9.2 is not a symmetrical device; the source and drain terminals are not interchangeable. The shallow n^+ source and drain regions are formed by ion implanting arsenic and the p-type region by implanting boron. A high temperature step activates the dopants and allows the boron, which diffuses more rapidly than arsenic, to diffuse and form the p-type body and channel of the n-channel MOSFET. The n-type region between the n^+ drain and the end of the p-type channel is an "extended drain," which lowers the electric field to increase the breakdown voltage. The thick oxide with the gate electrode on top is a *field-plate* that increases the breakdown voltage.

Figure 9.2 illustrates a fundamental trade-off. Longer drain extensions give higher drain breakdown voltages, which is desirable, but longer drain extensions give higher on-resistance, which is not desirable.

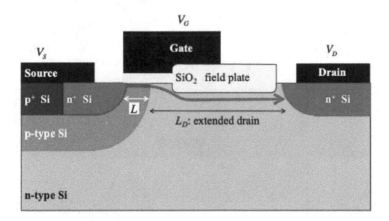

Fig. 9.2 Cross-section of a simple LDMOS transistor. (After [3].) The notation, n^+ and p^+, is used to indicate heavily doped n-type and p-type regions. The arrows indicate the direction that electrons flow from the source to the drain.

The heavily doped p^+ layer shown in Fig. 9.2 is used to make ohmic contact to the p-type body. Note that the source metallization covers both the n-type source and the p-type body contact. The source to body voltage is zero, so there is no body effect on the threshold voltage.

The vertical DMOSFET shown in Fig. 9.3 is a discrete device [2]. The source and channel of this transistor are similar to those in the LDMOS transistor and produced in the same way — the faster diffusion of the p-type dopants during the high temperature activation step produces the p-type channel. In this device, however, the drain contact is on the bottom. The arrows show the direction of electron flow when the device is on. In this device, the drain current flows vertically, while in the LDMOS transistor, the current flows laterally. As for the LDMOS transistor, the longer the drift region, the high the blocking voltage, but the on-resistance increases as well. Discrete, vertical devices of this type would have a blocking voltage on the order of 200–300 V or so.

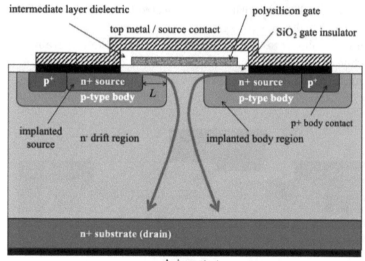

Fig. 9.3   Cross-section of a simple vertical DMOSFET [2]. The source and channel are similar to that of the LDMOS transistor, but the drain current flows vertically rather than laterally as in the LDMOS transistor. The arrows indicate the direction of electron flow from the source to the drain. Note that the figure is not drawn to scale. The n^- drift region is much longer that the junction depths of the n^+ source and p-type body.

Power MOSFETs operate at high currents and voltages. Large currents require a large width, W, of the MOSFET. Figure 9.3 is actually a cross section of one "cell" of the MOSFET. Each cell is a square (or hexagon in some designs) with the source along the outside edge and the current flowing down from the center of the square to the bottom of the wafer. A DMOSFET of a given area, A, consists of many of these cells connected in parallel in order to produce as large a MOSFET width as possible within a given area. The geometry of the DMOSFET makes it easy to develop expressions for R_{on} and V_{br}, and the specific results for this device illustrate a general trade-off between on-resistance and breakdown voltage.

9.3 On-resistance / blocking voltage trade-off

Figure 9.4 illustrates the components of R_{on} for a vertical DMOSFET. Beginning at the top, R_s is the series resistance of the source, which also includes the metal to source contact resistance. (See Sec. 4.6 for a discussion of series resistance.) The resistor, R_{ch}, is the channel resistance of the MOSFET, which was given by Eq. (4.14). The next resistance, R_{JFET} is the resistance as the current flows in the n-type region between the two p-type body regions. The depletion regions around the two pn junctions reduce the size of the region through which current flows. The term "JFET" stands for Junction Field-Effect Transistor, a transistor that operates by controlling the depletion layer width of a pn junction (see Chapter 15, Sec. 2 of [4]). The next resistance component is R_{dr}, the resistance of the n-type electron drift layer that is used to increase the drain breakdown voltage. Finally, there is R_{sub}, the resistance of the heavily doped n^+ layer used to make ohmic contact to the n-type drift layer.

For high voltage MOSFETs, the lightly doped n^- drift region must be long to support the high voltage without breaking down. As a result, $R_{on} \approx R_{dr}$, so

$$R_{on} \approx R_{dr} = \rho_{dr}\frac{t_{dr}}{A}, \tag{9.1}$$

where ρ_{dr} is the resistivity of the drift layer, t_{dr} is the thickness of the drift layer, and A is the cross-sectional area through which the current flows. The resistance scales as one over area, so it is common practice to quote the *specific on-resistance*, $R_{on,sp} = R_{on}A$, which has units of $\Omega - \mathrm{m}^2$. For the DMOSFET,

$$R_{on,sp} = R_{on}A \approx \rho_{dr}t_{dr} = \frac{t_{dr}}{N_D q \mu_n}, \tag{9.2}$$

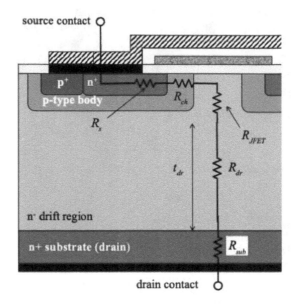

Fig. 9.4   Cross-section of a simple, vertical DMOSFET in the on-state showing the various components of R_{on}. Note that the figure is not drawn to scale. The n^- drift region is much longer that the junction depths of the n^+ source and p-type body.

where N_D is the doping density of the n^- drift layer, and μ_n is the electron mobility. A low on-resistance requires a high doping density and a short drift layer, but we will see that a high breakdown voltage requires a low doping density and a long drift layer, so there is an inherent trade-off between on-resistance and breakdown voltage.

Figure 9.5 shows the device for $V_{GS} < V_T$ where only leakage currents flow. A high voltage, V_{DS}, has been applied to the drain, and the source voltage is zero. The region in the dashed line rectangle can be approximated as a reverse biased, 1D pn junction with a depletion width, W_D, that extends mostly into the lightly doped n^- drift layer.

This is a problem that we can solve with the depletion approximation, as we did for the bulk MOSFET in Sec. 8.4. The electric field profile will look like that sketched in Fig. 8.7; the peak electric field, $\mathcal{E}_p$, occurs at the pn junction and is analogous to $\mathcal{E}_s$ in Fig. 8.7. The potential drop across the depletion region is $V_{DS} + V_{bi} \approx V_{DS}$, where $V_{DS} \gg V_{bi}$, which is the built-in potential of the pn junction. The depletion layer width, W_D, is related to the potential drop across the semiconductor according to

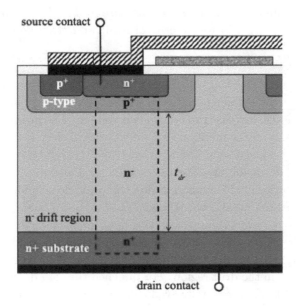

Fig. 9.5 Cross-section of a simple, vertical DMOSFET in the off state. The region in dashed line rectangle can be approximated as a one-dimensional, reverse biased pn junction. The p-layer is shown as p^+ to indicate that it (and the n^+ substrate) are much more heavily doped than the n^- drift layer. When the peak electric field at the junction reaches the critical field, avalanche breakdown occurs.

Eq. (8.8). Replacing ψ_s with $V_{DS} + V_{bi} \approx V_{DS}$, we find

$$W_D = \sqrt{2\kappa_s\epsilon_0 V_{DS}/(qN_D)}\,, \tag{9.3}$$

where we are assuming that $t_{dr} \geq W_D$; the n-type layer is thicker than the depletion layer.

The potential drop across the semiconductor (approximately V_{DS}) is the area under the $\mathcal{E}$ vs. x curve of Fig. 8.7, which gives

$$V_{DS} \approx \frac{1}{2}\mathcal{E}_p W_D\,, \tag{9.4}$$

where $\mathcal{E}_p$ is the peak electric field (analogous to $\mathcal{E}_s$ in Eq. (8.7)). Equations (9.4) and (9.3) are two equations for W_D, which can be solved to find

$$V_{DS} = \left(\kappa_s\epsilon_0\mathcal{E}_p^2\right)/(2qN_D)\,. \tag{9.5}$$

When the peak electric field reaches a material (and doping) dependent *critical field*, $\mathcal{E}_c$, *avalanche breakdown* occurs [4], At this electric field, electrons have gained sufficient energy to break covalent bonds and create electron-hole pairs (so-called *impact ionization* [4]). By setting the peak electric field in Eq. (9.5) to the critical field, we obtain an expression for the breakdown voltage,

$$V_{br} = \left(\kappa_s \epsilon_0 \mathcal{E}_c^2 \right) / (2qN_D) . \tag{9.6}$$

As expected, lighter doping densities, which reduce the peak electric field for a given voltage, increase the breakdown voltage. The breakdown voltage strongly depends on the material-dependent critical electric field for impact ionization. For a given semiconductor such as Si, careful device design can raise the breakdown voltage. The use of a lightly doped drain (the extended drain or drift layer in the figures) lowers the peak electric for a given V_{DS}, which increases the breakdown voltage, V_{br}.

Note that we have assumed that $W_D \leq t_{dr}$ in these calculations. If the doping density is very low, or t_{dr} is too thin, then the device will *punch through*. When N_D is very low, Eq. (8.5) tells us that the electric field is nearly constant in the drift region. For punched through devices, V_{br} is not simply related to $\mathcal{E}_c$; a more involved calculation of the ionization integral is required (See Sec. 2.2.6 in [2]).

Equation (9.6) is an upper limit for the breakdown voltage because real devices are three-dimensional, and there is *electric field crowding* at the edges of the junctions. Recall that $\vec{\mathcal{E}} = -\nabla V$, so the curvature at the edges of the junctions concentrates the electric field. Sophisticated *edge termination techniques* are used to achieve breakdown voltages that approach that of Eq. (9.6). The field plate in Fig. 9.2 is an example. (See Sec. 10.1.2 in [2] for a discussion of edge termination techniques.) It is also important to note that the gate oxide can break down. Careful design is required to keep the electric field in the oxide low enough so that breakdown is limited by the semiconductor. For Si power devices, the critical field for avalanche breakdown in the Si sets the breakdown voltage, but for wide band gap semiconductors for which the critical field is much higher, oxide breakdown can occur. High voltages can also reduce the long term reliability of the gate oxide, even if breakdown does not occur.

Now let's relate the specific on-resistance, Eq. (9.2), to the breakdown voltage. Equation (9.6) relates the breakdown voltage to the doping density; solving for the doping density, we find

$$N_D = \kappa_s \epsilon_0 \mathcal{E}_c^2 / (2qV_{br}) . \tag{9.7}$$

Equation (9.4) relates the voltage to the width of the depletion layer. Let's assume a device that is designed to be on the edge of punchthrough, $W_D =$

t_{dr} when the breakdown voltage is reached. Solving Eq. (9.4) for the thickness of the drift layer, we find

$$t_{dr} = W_D = 2V_{br}/\mathcal{E}_c. \qquad (9.8)$$

Finally, let's return to the specific on-resistance. Equation (9.2) gives $R_{on,sp}$ in terms of t_{dr} and N_D. Using Eq. (9.8) for t_{dr} and Eq. (9.7) for N_D, we find from Eq. (9.2)

$$\boxed{R_{on,sp} = \frac{4V_{br}^2}{\mu_n \kappa_s \epsilon_0 \mathcal{E}_c^3}.} \qquad (9.9)$$

Equation (9.9) is an important result that summarizes the trade-off between on-resistance and breakdown voltage. A high breakdown voltage (desirable), gives a high on-resistance (undesirable), and a low on-resistance (desirable) gives a low breakdown voltage (undesirable). But Eq. (9.9) also shows the strong role of the critical electric field for avalanche breakdown. For a semiconductor with a high critical field, we can achieve much lower on-resistance for a given blocking voltage. It is the job of the device designer to minimize the on-resistance for the required blocking voltage.

Device design consists of specifying the required drift layer thickness, t_{dr} and doping density, N_D, for the given blocking voltage, V_{br}, and required on-resistance, $R_{on,sp}$. In doing these calculations, it is important to understand that the mobility (important for on-resistance) depends on the doping density and so does the critical electric field. As mentioned earlier, the critical field also depends on the device structure — whether the device is punched through or not (see Sec. 2.2.6 in [2]). In deriving Eq. (9.9), we have assumed a device that is not punched through with $W_D = t_{dr}$. For a punched through device with $W_D > t_{dr}$, the numerical factor, 4, in Eq. (9.9) changes [2]. It turns out that optimally designed devices are slightly punched through [2].

The critical electric field for avalanche breakdown controls the $R_{on,sp}$ vs. V_{br} trade-off. It increases with band gap because for a larger band gap, it takes more energy to move an electron from the valence to conduction band. Silicon has a band gap of 1.1 eV and moderately doped Si ($N_D \approx 10^{16}$ cm^{-3}) has a critical electric field of about 0.4 MV/cm. Silicon carbide comes in different polytypes; 4H-SiC has a band gap of about 3.3 eV and a critical field of about 2.3 MV/cm at $N_D = 10^{16}$ cm^{-3}. Gallium nitride has a similar band gap (≈ 3.4 eV) and a similar critical electric field [6].

Figure 9.6 compares the upper limit on-resistance vs. blocking voltage as given by Eq. (9.9) for Si, SiC, and GaN. Assume that we need a power

MOSFET with a blocking voltage of 1,000 V. Figure 9.6 shows that a Si power MOSFET would have a specific on-resistance of about 300 $m\Omega - cm^2$ while a wide band gap semiconductor would be about 0.3 $m\Omega - cm^2$. To achieve a particular on-resistance in Ohms ($R_{on} = R_{on,sp}/A$), the Si device would need to be 1000X larger than a wide wide band gap device. The cost of a device is proportional to its area, so the wide band gap device would be significantly cheaper.

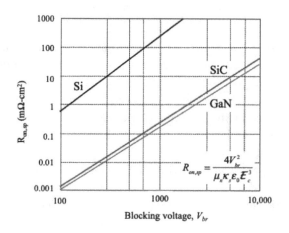

Fig. 9.6 On-resistance vs. blocking voltage for Si, 4H-SiC, and GaN as given by Eq. (9.9). Note that because the critical electric field is difficult to measure, there is some uncertainty in the lines shown above, but they clearly illustrate the motivation for developing wide band gap power devices. (After [6].)

9.4   Figure of Merit (FOM)

A figure of merit (FOM) is a single number used to compare power devices. As shown in Fig. 9.1, the on-current, I_{on}, and the blocking voltage, V_{br}, are two key parameters, so we can define an FOM as [2]

$$FOM = \frac{I_{on}V_{br}}{A} = J_{on}V_{br}\,, \tag{9.10}$$

where A is the area of the device and J_{on} is the current density. The higher the FOM, the better the device.

The power dissipated by the device heats it up. The power dissipated is

$$P_D = I_{on}V_{on}\left(\delta t/T\right) + I_{off}V_{off}\left(1 - \delta t/T\right) + P_{sw}\,, \tag{9.11}$$

where $\delta t/T$ is the *duty cycle* (the fraction of the time that the MOSFET is on), and P_{sw} is the power dissipated while switching. Equation (9.11) can be compared to the power dissipation of a CMOS gate as discussed in Sec. 2.2. In the CMOS inverter, current only flows while switching, so $I_{on} \approx 0$. There is off-state leakage, but it only becomes important for chips with millions or billions of transistors. For CMOS inverters, we focussed on the dynamic power, P_{sw} in Eq. (9.11).

For Si power MOSFETs, we can ignore the off-state power dissipation, and when the switching frequency is low, the switching power can also be ignored, so we will focus for now on the first term in Eq. (9.11).

Assuming a 100% duty cycle, we find the power density to be

$$P_D/A = \frac{I_{on}V_{on}}{A} = \frac{I_{on}^2 R_{on}}{A} = J_{on}^2(R_{on}A) = J_{on}^2 R_{on,sp} . \qquad (9.12)$$

The device temperature is proportional to the power dissipated:

$$\Delta T = P_D \theta_{th} , \qquad (9.13)$$

where θ_{th} is the *thermal resistance*. Equation (9.13) is analogous to $V = IR$ with the temperature rise analogous to voltage, the power dissipated analogous to current, and the thermal resistance analogous to the electrical resistance. The thermal resistance consists of an intrinsic component that depends on the device geometry and the thermal conductivity of the semiconductor, a component for the semiconductor to case thermal resistance, and a component for the case to ambient thermal resistance. Operation of devices at high powers requires careful attention to thermal design.

There is a maximum junction temperature, $T_{j,max}$, at which the device fails. The corresponding maximum power is

$$P_{max}A = \frac{T_{j,max} - T_{amb}}{\theta_{th}} = I_D V_{DS} , \qquad (9.14)$$

and from Eq. (9.12), the corresponding current density is

$$J_{on} = \sqrt{P_{max}/R_{on,sp}} . \qquad (9.15)$$

Finally, using this result in the FOM, Eq. (9.10), we find [2]

$$\boxed{FOM = J_{on}V_{br} = \sqrt{P_{max}\left(V_{br}^2/R_{on,sp}\right)} .} \qquad (9.16)$$

The maximum power density in the FOM is limited by the material dependent $T_{j,max}$, the thermal conductivity of the semiconductor, and by the thermal design of the package. The goal of the thermal designer is to minimize the temperature rise for a given power dissipation. The factor, $(V_{br}^2/R_{on,sp})$, in the FOM is known as the *unipolar device FOM* [2]. The goal of the device designer is to maximize this factor. When the device is designed to be on the edge of punchthrough, $(V_{br}^2/R_{on,sp})$ depends only on material parameters, and Eq. (9.9) can be used to define a material figure of merit as

$$BFOM = \frac{V_{br}^2}{R_{on,sp}} \propto \mu_n \kappa_s \epsilon_0 \mathcal{E}_c^3 . \tag{9.17}$$

The factor of four in Eq. (9.9) has been dropped because the precise numerical value depends on whether the device is punched through or not.

This particular FOM, the *Baliga figure of merit*, allows us to compare the power semiconductor device performance potential of different semiconductors. The higher the BFOM, the better. Comparing BFOM's is a simple way to assess the potential of different semiconductors for use in power electronics, but this approach does not capture all of the important effects. For example, the thermal conductivity of SiC is nearly three times higher than that of GaN, which gives SiC an advantage in terms of self-heating. The electron mobility in the channel of a GaN FET is higher than for SiC, which gives GaN an advantage in low blocking voltage applications for which $R_{ch} > R_{dr}$. There are, in fact several different FOMs that capture different aspects of device performance, and they can give inconsistent predictions of which material provides better power device performance [5]. The Safe Operating Area (SOA) of a power MOSFET integrates factors considered by different FOMs provides a more comprehensive way to compare technologies [5].

9.5   Safe Operating Area (SOA)

As shown in Fig. 9.7, the Safe Operating Area (SOA) of a power MOSFET defines the voltage and current limits within which the device can safely operate. An SOA is often provided on transistor data sheets. The SOA is determined by four factors: 1) on-resistance, 2) maximum current permitted, 3) maximum power dissipation permitted, and 4) breakdown voltage. Figure 9.7 shows two cases: 1) DC operation or 100% duty cycle and 2) pulsed operation with a duty cycle less than 100%.

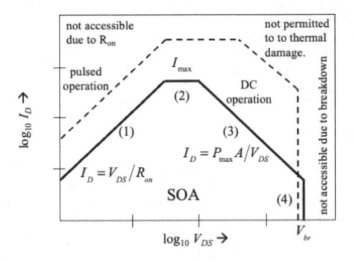

Fig. 9.7 Illustration of the Safe Operating Area of a power transistor. Two cases are shown: 1) DC operation (solid line) and 2) pulsed operation (dashed line).

For low V_{DS}, the current is limited by the on-resistance of the MOSFET; $I_D = V_{DS}/R_{on}$, which is line (1) with a slope of +1 on the log-log plot in Fig. 9.7. Line (1) is lower for DC operation than for pulsed operation because self-heating, more pronounced under DC conditions, increases the temperature, which lowers the mobility and increases R_{on}, which lowers I_D. As also shown in Fig. 9.7, there is a maximum current, I_{max}, at which the wire leads that connect the transistor to the pins of the package fail. This is line (2) in Fig. 9.7. As the voltage increases along line (2), the power dissipated increases until the maximum power density, P_{max} is reached. Line (3) of the SOA is determined by the *self-heating* that occurs when significant power is dissipated. From Eq. (9.14), we find $I_D = P_{max}A/V_{DS}$, which is line (3) in Fig. 9.7, a line with a slope of -1. (It should be noted that the slope of line (3) can be steeper than -1 if thermal instabilities occur.) Finally, when V_{DS} reaches V_{br}, we reach line (4), which is the right boundary of the SOA. The avalanche breakdown voltage increases with temperature. The temperature dependence is small, but V_{br} is slightly smaller under pulsed operation than for DC operation because self-heating is less under pulsed operation.

When MOSFETs are used in power electronics, they are typically switched on and off at a frequency, f, with the fraction of the time on, the duty cycle, $\delta t/T$, being less than 100%. As shown in Fig. 9.7, for a duty cycle of less than 100%, higher instantaneous currents and power dissipations can be tolerated. The designer must ensure that I_D and V_{DS} stay within the SOA at all times, even while switching during which capacitances and inductances can produce voltage and current transients. Switching power is discussed in the next section.

9.6 Switching power

Power dissipation sets the energy consumption of power electronics circuits and produces the self-heating that can can cause devices to fail. As described by Eq. (9.11), power is consumed when the device is on, when it is off, and while it is switching. A major reason that MOSFETs are used is because they can be operated at higher frequencies than IGBTs [1], but at high frequencies, the switching power in Eq. (9.11) becomes significant.

The switching power is analogous to the dynamic power of a CMOS inverter (see. Eq. (2.6) in Sec. 2.2). For a CMOS inverter, we wrote

$$P_{sw} = \alpha f C_{sw} V_{DD}^2 \,, \tag{9.18}$$

where α, the activity factor, is the probability that a gate will change state on a given clock cycle. Equation (9.18) describes the energy dissipated in charging and discharging the switching capacitance, C_{sw}. For a CMOS inverter the current is approximately zero when $V_{in} = 0$ and when $V_{in} = V_{DD}$, but if $V_{in} = V_{DD}/2$, current will flow in both the NMOS and PMOS transistors, and power will be dissipated. This *short-circuit power* occurs whenever there are finite rise and fall times to the input voltage, but it is not the primary source of power dissipation in CMOS gates. Power MOSFET circuits, however, dissipate power then they are on and when they are switching. At high frequencies, this switching power can dominate.

Figure 9.8 is a simple power MOSFET switching circuit. The inductor may represent the windings of a motor that is being controlled. The diode provides a path for current to flow when the MOSFET is off. The inductance is assumed to be large enough to keep the current nearly constant during the switching transient. Let's compute the power dissipated during the turn on transient when the gate voltage is stepped up. To keep the discussion simple, we'll assume that the power dissipated during the turn off transient is the same.

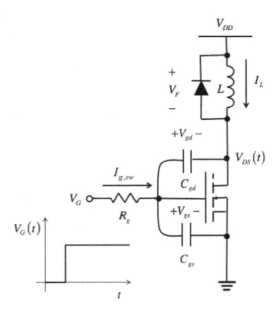

Fig. 9.8 Simple MOSFET switching circuit showing the device capacitances, C_{gs}, and C_{gd}, and the inductive load. Only the turn on transient will be discussed.

The switching power is dominated by C_{gd} because the *Miller effect* of the circuit increases the effect of C_{gd}. For the turn on transient, the device is switched from off to on, the change in voltage across C_{gd} is

$$\Delta V = V_{\text{off}} - V_{\text{on}} = V_{DD} + V_F - V_{\text{on}} \approx V_{DD}. \tag{9.19}$$

The corresponding change in charge is

$$Q_{gd} \approx \overline{C_{gd}}\, V_{DD}, \tag{9.20}$$

where an average C_{gd} must be used because it varies strongly with voltage. This charge is supplied by the gate current, $I_{g,sw}$, during the switching transient. We can solve for the corresponding time, t_{sw}, from

$$I_{g,sw}\, t_{sw} = \overline{C_{gd}}\, V_{DD} \rightarrow t_{sw} = \frac{\overline{C_{gd}}\, V_{DD}}{I_{g,sw}}. \tag{9.21}$$

The average power dissipation is the switching energy, E_{sw}, divided by the period of the clock, $T = 1/f$. The average power dissipated during the turn-on transient is

$$P_{sw} = \frac{E_{sw}}{T/2} = 2fE_{sw} = 2f\left[(I_{on}V_{DD}/2)\right]t_{sw}, \tag{9.22}$$

where we have assumed that the inductor holds the current nearly constant, so $I_D \approx I_{on}$ during the transient while V_D drops from $\approx V_{DD}$ to $V_{on} \approx 0$. The term in square brackets is the average power dissipated during the charging transient. Finally, using Eq. (9.21) for the charging time, we find

$$P_{sw} = \beta f \overline{C_{gd}} V_{DD}^2 = \alpha f Q_{gd} V_{DD}, \qquad (9.23)$$

where

$$\beta \equiv I_{on}/I_{g,sw} . \qquad (9.24)$$

Equation (9.23) for the power MOSFET switch is identical in form to Eq. (9.18) for the CMOS inverter, but the physics is different. Data on Q_{gd} rather than $\overline{C_{gd}}$ is typically provided on MOSFET data sheets.

The average power dissipation is proportional to frequency, capacitance, and voltage squared, as expected. We can write the switching power as a function of several variables:

$$P_{sw} = \text{function}\left(R_g, I_{on}, \overline{C_{gd}}, V_{DD}, f\right) . \qquad (9.25)$$

The factor, β, should be minimized. A larger gate switching current, $I_{g,sw}$, reduces α — reducing the average power dissipation by shortening the switching time, t_{sw}. A smaller R_g, increases the gate drive current, $I_{g,sw}$, and reduces the power. A larger I_{on} increases the power dissipation during switching. The capacitance can be minimized by careful device design. For example, the thick oxide in the field plate of Fig. 9.2 is needed to keep C_{gs} small. The voltage is set by the application, and the maximum frequency by the maximum power permitted by the device and the application.

We've discussed two power MOSFET figures of merit as given by Eqs. (9.16) and (9.17). Another widely-used figure of merit is

$$FOM = R_{on}Q_{gd} . \qquad (9.26)$$

In this FOM, R_{on} controls the power dissipation while the MOSFET is on, and Q_{gd} determines the power dissipation while the device is switching. Information about both is typically provided on power MOSFET data sheets. For a thorough discussion of switching in power MOSFETs, see Secs. 6.13–6.16 of [1] , and for an analysis of switching times, see Sec. 8.2.12 in [2].

9.7   Discussion

It is important to understand that because power MOSFETs are operated under high current/high voltage conditions, a number of second order effects can occur. For example, a parasitic device can turn on. Note that the n^+ source, p-type body, and n-type drift layer in the DMOSFET of Fig. 9.3 form a parasitic npn bipolar junction transistor (BJT). (See Sec. 10.5 for an introduction to BJTs.) If this parasitic bipolar transistor turns on, it cannot be turned off because the gate has no control over it. This phenomenon can produce catastrophic failure of the device.

The parasitic BJT can turn on when high drain voltages produce high electric fields, which create electron-hole pairs by impact ionization. High device operating temperatures, can also create electron-hole pairs by thermal generation. Holes collected by the p-type base represent a base current that can turn the BJT on. Rapid switching can also produce capacitive transient base currents the turn the BJT on. Careful device design guided by an understanding of device physics is needed to insure that the parasitic BJT does not turn on while the transistor is in its safe operating region.

9.8 Summary

This lecture has been a short introduction to power MOSFETs and to some considerations relevant to all power semiconductor devices. There are several variations on the two MOSFET designs we discussed, and in addition to MOSFETs, bipolar transistors and insulated gate bipolar transistors (IGBTs) are also used. Power MOSFETs can be CMOS compatible and used along with digital electronics for mixed signal integrated circuits, or they can be discrete devices used to control motors, relays, lights, etc.

When used for RF power, a power device should have a high transconductance, a high output resistance, and a high breakdown voltage. When used as a switch, a power device should have a low on-resistance and high breakdown (or blocking) voltage. We have seen that there is an inherent trade-off between achieving low on-resistance and a high blocking voltage. This trade-off is summarized by the figure of merit. For a growing number of applications, wide band gap semiconductors can out perform silicon.

This lecture is intended to be a starting point. For a more in depth treatment of power MOSFETs and a more comprehensive treatment of power semiconductor devices, see [1, 2].

Lecture 9 Exercise: Wide band gap semiconductors

Consider a power electronics application for which a blocking voltage of 1,000 V and an on-resistance of 20×10^{-3} Ω is needed. If we choose a non-punched-through Si power MOSFET, what is the area of the MOS-FET? Some relevant material parameters for a blocking layer with $N_D = 10^{16}$ cm^{-3} are

$$\mathcal{E}_c = 0.4 \times 10^6 \ \text{V/cm}$$

$$\epsilon_s = \kappa_s \epsilon_0 = 11.8 \epsilon_0$$

$$\mu_n = 980 \ \text{cm}^2/\text{V} - \text{s}.$$

From Eq. (9.9), we find

$$A = \frac{4 V_{br}^2}{(\mu_n \kappa_s \epsilon_0 \mathcal{E}_c^3) R_{on}}, \qquad (9.27)$$

which gives

$$A = \frac{4 \times (10^3)^2}{(980 \times (11.8 \times 8.854 \times 10^{-14}) \times (0.4 \times 10^6)^3 \times (20 \times 10^{-3})},$$

or

$$A \approx 3 \ \text{cm}^2.$$

If, on the other and, we choose a SiC power MOSFET, what is the area of the device? Some relevant material parameters for a blocking layer with $N_D = 10^{16}$ cm^{-3} are

$$\mathcal{E}_c = 3.3 \times 10^6 \ \text{V/cm}$$

$$\epsilon_s = \kappa_s \epsilon_0 = 9.7 \epsilon_0$$

$$\mu_n = 850 \ \text{cm}^2/\text{V} - \text{s}.$$

Again, using Eq. (9.9), we find

$$A = \frac{4 \times (10^3)^2}{(850 \times (9.7 \times 8.854 \times 10^{-14}) \times (3.3 \times 10^6)^3 \times (20 \times 10^{-3})},$$

or

$$A \approx 0.07 \ \text{cm}^2.$$

The much smaller SiC device would make it much less expensive for this application.

9.9 References

For comprehensive treatments of power semiconductor devices, consult the following books. The second one treats Si power devices as well SiC devices.

[1] B. Jayant Baliga, *Fundamentals of Power Semiconductor Devices*, 2nd Ed., Springer, 2019.

[2] T. Kimoto and J.A. Cooper, *Fundamentals of Silicon Carbide Technology: Growth, Characterization, Devices, and Applications*, Wiley-IEEE Press, 2014.

For an analysis of a representative LDMOS transistor, see:

[3] Y. Chen, B.K. Mahajan, D. Varghese, S. Krishnan, V. Readdy, and M.A. Alam, "A novel 'I-V Spectroscopy' technique to deconvolve threshold voltage and mobility degradation in LDMOS transistors," *IEEE International Reliability Physics Symposium*, pp. 1-6, 2020.

The JFET is discussed in Chapter 15, Sec. 2 of the following book. Avalanche breakdown is discussed in Chapter 6, Sec. 6.2.2.

[4] R. F. Pierret, *Semiconductor Device Fundamentals*, Addison-Wesley Publishing Co., Reading MA, 1996.

As discussed in the following paper, the Safe Operating Area (SOA) can be used to compare different power semiconductor device technologies.

[5] B.K. Mahajan, Y.-P. Chen, N. Zagni, and M.A. Alam, "Self-heating and reliability-aware 'Intrinsic' safe operating area of wide bandgap semiconductors – An analytical approach," *IEEE Trans. on Device and Materials Reliability*, **21**, pp. 518-527, 2021.

Recent data on the critical electric field and on-resistance/blocking voltage trade-off in SiC and GaN are discussed in the following paper.

[6] J.A. Cooper and D.T. Morisette, "Performance Limits of Vertical Unipolar Power Devices in GaN and 4H-SiC," *IEEE Electron Device Letters*, **41**, pp. 892-895, 2020.

Lecture 10

Transistors and Semiconductor Memory

10.1 Introduction

Digital information processing requires both logic and memory. Logic is performed with transistors, and much of the memory is implemented with transistors too. This lecture is a short introduction to the use of transistors in semiconductor memories. Memories are classified in different ways. Some are *serial*, in which one must wait for the data of interest to come out in a stream of data. In *Random Access Memory* (RAM), any arbitrary memory location may be accessed. Memories can be *volatile* or *nonvolatile*. A volatile memory loses its information when the power is switched off, but a nonvolatile memory retains its information (at least for several years). Two types of nonvolatile memory are mask-programmed *Read Only Memory* (ROM) and *electrically erasable programmable* read only memory (EEPROM). Flash memory is a type of EEPROM. Other types of nonvolatile memory are resistive RAM (ReRAM), phase change memory (PCM), ferroelectric RAMs (FeRAMs), and magnetic RAMs (MRAMs).

The *access time* of a memory is the time between the start of a read operation and when the data appears on the output port. The memory *cycle time* is the minimum time time between two memory operations, which is the sum of the time to read from and write to the same location in memory. Depending on the type of memory, the cycle time may range from few nanoseconds to a few hundred nanoseconds. The trade-off between access and cycle time and the storage capacity of a memory technology leads to the use of a memory hierarchy as shown in Fig. 10.1.

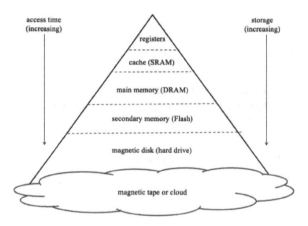

Fig. 10.1   Illustration of the memory hierarchy, which is used in computer systems to utilize very fast memory, which provide small amounts of data storage and very large storage capacity memory, which provides data at a slower rate.

Central processing units (CPUs) run very fast, and to keep up, data must be quickly transferred into and out of memory. The registers in the CPU provide very fast but a very limited amount of memory. *Static Random Access Memory* (SRAM) provides very fast on-chip memory, but it takes up valuable real estate on the chip, so the amount of on-chip SRAM that can be provided is limited. On-chip SRAM is called *cache* memory; its job is to keep the CPU supplied with data. The access time for SRAMs is on the order of a nanosecond.

Data is moved in and out of the cache from off-chip memory, typically a type of volatile memory known as *Dynamic Random Access Memory* (DRAM), which resides on other chips. The wires that transmit the data off and on chip consume power and increase the access time. The access time for DRAMs is currently about 35 nanoseconds, much longer than for

SRAMs. Processor speeds are much faster than DRAM access times, which has led to a *memory bottleneck* that limits the performance of computing systems.

Data that must be stored indefinitely, even when the chips are switched off (e.g. music and video files), is stored on so-called *flash* memory chips. Their cycle time is much longer than for SRAMs and DRAMs, but the storage capacity is large. When very large amounts of data must be stored, magnetic hard drives or magnetic tapes are used. These are serial memories with long cycle times.

Our focus in this lecture is on individual memory cells implemented with transistors. These memory cells are organized into arrays. Before we discuss memory cells, we take a quick look at memory chips.

10.2 Memory chips

Figure 10.2 shows how semiconductor memories are organized. The core of the chip is an array of *memory cells* each of which is capable of storing one (or sometimes more) bits of information. A memory cell may consist of from one to six transistors or even 10 transistors for radiation hard circuits. Each cell is connected to a row, known as the *wordline* and to a column known as the *bitline*. The wordlines are connected to the gates of transistors and the bitlines to the drains. To read from or write to a particular cell, its word and bit lines are selected. Single alphanumeric characters are represented by 8-bits, one *byte*. Scientific computing systems are organized in terms of *words*, which may be 16, 32, 64, or 128 bits. Memories may simultaneously select a bit, a byte, or a word of data. Sometimes, larger groups of data, strings, pages, or blocks, are accessed at once.

In a random access memory, a *read operation* begins with an M-bit row address, which the *row decoder* uses to select a specific wordline and raise its voltage. All cells in the row then provide their data to their bit lines. For example, if the cell is storing a logic 1 (a high voltage), then the voltage of that cell's bit line will increase a little. The small readout signal is amplified by the *sense amplifier*. The output signals of all cells in the wordline are delivered to the *column decoder*, which selects the column specified by an N-bit address and sends it to the input/output (I/O) data line.

The *write operation* begins by supplying the data to be stored to the I/O line. The specific cell to which the data is stored is specified by an N-bit column address and an M-bit row address. In this case, the sense amplifier acts as a *driver* and writes the data to the memory cell. Large memory arrays are divided into smaller blocks, which makes it easier to get data into and out of the memory cells.

Our focus in this lecture is on how transistors are used to realize different types of memory cells. For more information on overall memory chip design, see [1].

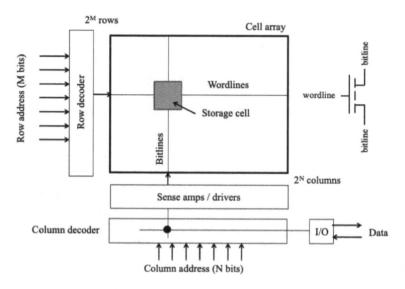

Fig. 10.2 Simple illustration showing how a memory chip is organized. Typically, the cell array occupies 60% of the area of a chip.

10.3 Static Random Access Memory (SRAM)

Figure 10.3 shows a 6-transistor SRAM memory cell. There are two bitlines; one is the complement of the other (i.e. if one bit line is a logic 1, the other is a logic zero and vice versa). When the voltage on the wordline is high, transistors M_5 and M_6 are turned on to connect the two bitlines to the memory cell. The memory cell itself consists of two cross-coupled inverters of the type discussed in Sec. 2.2. Designers work hard to minimize the area and power consumption of the cell so that as many cells as possible can be placed on the chip, close to the CPU.

To understand the operation of the SRAM cell, consider what happens if the voltage, V_1, is high (a logic 1). This means that transistor, M_1, is off.

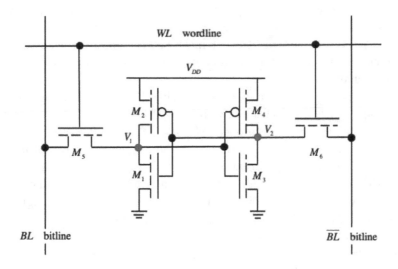

Fig. 10.3   A 6-transistor SRAM memory cell.

The voltage, V_1, is connected to the input of the second inverter, so if V_1 is high, transistor M_3 will be turned on and M_4 will be off, which makes $V_2 = 0$ V, a logic zero. Since V_2 is cross-connected to the input of the first inverter, $V_2 = 0$ will ensure that M_1 is, indeed, off and M_2 is on. The state, $V_1 = 1, V_2 = 0$ is a stable state of the memory cell. By similar arguments, if we assume $V_1 = 0$, we see that $V_1 = 0, V_2 = 1$ is also a possible state; these are the two stable states of this memory cell.

For a symmetric cell, there is also a possible state at $V_1 = V_2 = V_{DD}/2$, but this is not a stable stable state. If the voltage, V_1 increases by δV due to a small noise perturbation, the corresponding increased voltage on the gates of the second inverter will lower V_2, which as the input to the first inverter will raise V_1 more. This is a positive feedback that will drive V_1 to V_{DD} or 0 depending on whether the perturbation is positive or negative. When reading the contents of the cell, V_1 and V_2 are perturbed, and the designer must take care that these perturbations don't reach the switching voltage, or the read operation will change the state of the cell. Because there are only two stable states for this circuit, we can use it to store one bit of information: a logic zero or a logic one.

Next, let's consider how data would be read from an SRAM cell. The two bitlines are first *precharged* to a high voltage of about V_{DD}. The wordline voltage is then raised to V_{DD}. Note that the wordline is connected to many memory cells in a row, so it has a large capacitance. Recall from the discussion in Sec. 2.2, that capacitance slows down voltage transitions (see Eq. (2.2)), so the address decoder must be designed to provide sufficient drive current to charge the wordline so that an acceptable memory access time is achieved.

For a specific read example, assume that $V_1 = 0$ V and $V_2 = 1$ V, so M_1 is on and M_2 is off when the wordline voltage is raised. Current begins to flow from the bitline, through M_5 and M_1 to ground. This current discharges the bitline capacitance, C_{BL}. The voltage on the complementary bitline, $\overline{BL}$, remains high because there is no path to ground through M_3. Note that a large current through M_1 would discharge the bitline quickly, but this would require a large transistor and the correspondingly large SRAM cell that would limit the number of bits that could be stored on the chip. Instead, small transistors are used, which produces a small difference in the voltages on BL and $\overline{BL}$ that the sense amplifier uses to generate a logic zero signal at the I/O port. When the read operation is complete, the wordline voltage is re-set to zero and the bitlines are pre-charged back to a high voltage. Finally, note that when the data is read, the current through M_1 causes its drain voltage, V_1, to increase a bit, which causes V_2 to decrease a bit. The SRAM cell must be designed so that these perturbations are small and don't flip the state of the cell. When designed properly, readout of the data is *nondestructive*.

We have described how we would read a logic zero at V_1; how would we write a logic one to V_1? To do this, we begin with both bitlines pre-charged to approximately V_{DD}, and then pull the $\overline{BL}$ voltage to zero. The wordline voltage is increased, which connects the high voltage on BL to the drain of M_1, node V_1 and the low voltage on $\overline{BL}$ to the drain of M_3, node V_2. If the state of the cell was $(V_1 = 0, V_2 = 1)$, this will cause the cell to switch state to $(V_1 = 1, V_2 = 0)$. If the cell was already in this state, nothing happens.

SRAM memories are fast, so they can keep up with the demands of a fast CPU, but the fact that six transistors are needed to store a single bit limits the density of the memory. SRAM memories are also volatile. If the chip loses power, the data is lost. Much higher densities of memory can be obtained with Dynamic Random Access Memory (DRAM), another type of volatile memory. As discussed next, DRAM technology is different from logic transistor technology, so DRAM and the CPU with SRAM are implemented on separate chips. Sending and retrieving data off chip to the DRAM adds delays and increases power. Memory management is about keeping the SRAM cache full of the data that the CPU will need soon.

10.4   Dynamic Random Access Memory (DRAM)

Figure 10.4 shows a DRAM memory cell with only one transistor, the *access transistor*, whose gate is connected to the wordline. Note also that there is only one bitline. When the storage capacitor, C_S, is charged to V_{DD}, it represents a logic one, and when it is discharged, it represents a logic zero. Because each DRAM cell has only one transistor, DRAM is capable of storing much more data per unit area than SRAM, which uses six transistors per cell. For current technology, the area of a DRAM cell is about five times smaller than the area of an SRAM cell.

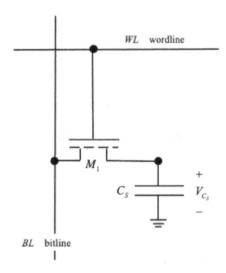

Fig. 10.4   A one transistor DRAM memory cell showing the storage capacitor, C_S and the access transistor. The bitline has a capacitance of $C_{BL} \gg C_S$ with the bitline voltage, V_{BL} across it.

To read data from a DRAM cell, the bitline is first precharged. The wordline voltage is then raised to V_{DD}, which connects every storage capacitor in the row to its bitline. When the access transistor is on, the storage capacitance, C_S, is connected in parallel with the bitline capacitance, C_{BL}. If the voltage across C_S is V_{C_S}, and the voltage across C_{BL} is V_{BL}, the precharge voltage, then before the wordline is activated, the total charge on these separate capacitors is $C_S V_{C_S} + C_{BL} V_{BL}$. After the wordline is selected, the voltage across the two capacitors in parallel must be such that the total charge is unchanged. Charge conservation gives

$$C_S V_{C_S} + C_{BL} V_{BL} = (C_S + C_B)(V_{BL} + \Delta V),\qquad(10.1)$$

where ΔV is the change in the bitline voltage. Solving for ΔV, we find

$$\Delta V = \frac{C_S}{C_S + C_{BL}}(V_{C_S} - V_{BL}) \approx \frac{C_S}{C_{BL}}(V_{C_S} - V_{BL}),\qquad(10.2)$$

where the approximation is made because $C_{BL} \gg C_{C_S}$; the long bitline has much more capacitance than the single storage capacitor. Now assume that the precharge voltage is $V_{BL} = V_{DD}/2$ and consider two cases: i) reading a logic one, $V_{C_S} = V_{DD}$ or ii) reading a logic zero, $V_{C_S} = 0$. We find that the change of voltage on the bitline is

$$
\begin{aligned}
\text{Case i)} \quad V_{C_S} &= V_{DD} & \Delta V &\approx +\frac{C_S}{C_{BL}}\left(\frac{V_{DD}}{2}\right) \\
\text{Case ii)} \quad V_{C_S} &= 0 & \Delta V &\approx -\frac{C_S}{C_{BL}}\left(\frac{V_{DD}}{2}\right),
\end{aligned}
\qquad(10.3)
$$

which is a very small change in voltage because $C_S \ll C_{BL}$. This small change in voltage is detected by the sense amplifier, which sends a logic one or zero to the I/O port. Note that the readout of the data is destructive because the voltage across C_S will not be 0 or V_{DD} after the read operation. In addition to sending the data to the I/O port, the sense amplifier also sends the data back to the all of the storage capacitors in the selected row thereby refreshing the data.

The write operation is similar. If a one is to be written, the bitline is precharged to V_{DD}, the wordline is activated, and C_S is charged to V_{DD}, but let's look more carefully at the charging process. If $V_{C_S} = 0$ initially, then the MOSFET lead connected to the bitline acts as the drain. The storage capacitor is connected to the source, and as the capacitor charges, the source voltage increases. If the gate voltage is V_{DD}, then the access transistor will turn off when $V_{C_S} = V_{DD} - V_T$, and the capacitor will not be fully charged to V_{DD}. To fully charge the capacitor, the wordline voltage is set at $V_{DD} + V_T$ during the write operation. Note also that if the access transistor is a bulk MOSFET, then there will be a body effect that increases V_T as the capacitor charges.

Like SRAM, DRAM is volatile; if the power is switched off, the data is lost, but unlike SRAM, even while powered on, a DRAM cell will "forget" its data in several milliseconds. The primary charge leakage mechanism is the subthreshold current of the access transistor. The data must be refreshed periodically, typically every 5–10 ms. This is what makes the memory "dynamic."

The DRAM storage capacitor is critical. A physically small but large enough capacitance is needed to provide a high density of memory with signals that are large enough to detect. Over the years there have been many innovations in capacitor design. Some DRAMs use capacitors above the substrate (so-called *stacked* or *folded plate* capacitors while others use capacitors below the substrate (so-called *trench* capacitors) [2]. The goal is to pack as large a capacitance as possible into as small an area as possible, but to achieve 16 gigabits per chip (current technology), the area must be small. Typical storage capacitances are about 10 femtofarads.

10.5   Read Only Memory (ROM)

In addition to volatile memories like SRAM and DRAM, nonvolatile memories to store data permanently are also needed. Read only memory (ROM) is a type of nonvolatile random access memory that provides a fixed binary output for every binary input. This type of memory, or *firmware*, is used to store program instructions, constants, and control information. Mask programmable ROMs are programmed with a fixed set of data when the chip is fabricated.

Two different types of ROM arrays are the NOR array and the NAND array. Figure 10.5 shows a NAND ROM array in which the output of each column is the logical NAND of all the word lines with transistors present. The presence of a transistor at a bitline/wordline crosspoint corresponds to a logical one and the absence of a transistor to a logical zero. This type of memory is produced by placing a transistor at each cross point and then ion implanting n-type dopants to short the source and drain — effectively removing the transistor at each location where a logical zero is to be stored.

To read the logical zero stored at (WL, BL) = (2, 2), BL_2 is selected by turning on the NMOS transistor at the top of the bitline. All wordline voltages except WL_2 are set to a high voltage, so the transistors at (1, 2) and (3, 2) are turned on. The BL_2 output voltage is low because there is a continuous path to ground along the BL_2. We have read the logical zero stored at (2, 2). To read the logical one stored by the presence of a transistor at (1, 3), BL_3 is selected and all wordlines except WL_1 are set to a high voltage. Transistor (1, 3) is off because the WL_1 voltage is low, which creates an open in BL_3 and forces the BL_3 output voltage high. We have read the logical 1 stored at (1, 3).

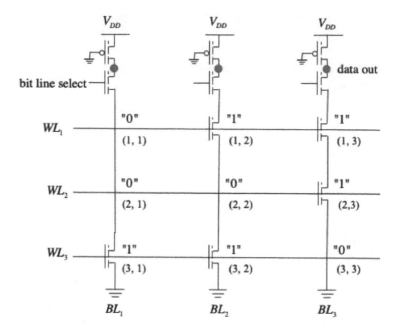

Fig. 10.5 Section of a mask programmable NAND ROM array showing three bitlines and three wordlines. Note that the PMOS transistors at the top are always on. (After Fig. 9.20 in [2].)

ROM memory arrays may be organized either as NOR arrays or as NAND arrays. NOR arrays usually have faster access times, but NAND arrays offer higher density. ROMs are programmed once, during chip fabrication, but for many applications, the ROM must be re-programmed from time to time. There are erasable and programmable ROMs (EPROMs) and electrically erasable and programmable ROMS, EEPROMs or E^2PROMs.

10.6   Floating gate memory cells

Flash memory is currently the dominant technology for the nonvolatile storage of large amounts of data. SRAM, DRAM, and ROM memory cells all make use of conventional transistors, but the Flash memory cell is a special transistor. As shown in Fig. 10.6a, the transistor has two gate oxide layers and two gate electrodes, but one of them is a *floating gate*. The floating gate electrode is heavily doped, n^+ polycrystalline silicon (polysilicon), and the top gate electrode may be polysilicon or metal.

To understand the operation of this transistor, consider the gate stack shown in Fig. 10.6c. When the transistor is programmed with a logical "1", there is a negative charge, Q_{FG}, on the floating gate. When the transistor is programmed with a logical "0", there is no charge on the floating gate. The solid lines in Fig. 10.6d show that when the control gate voltage (the wordline voltage) is set to V_R, then if the transistor is in its "0" state with $Q_{FG} = 0$, a significant drain current will flow, but if the transistor is in its "1" state, then little or no current flows.

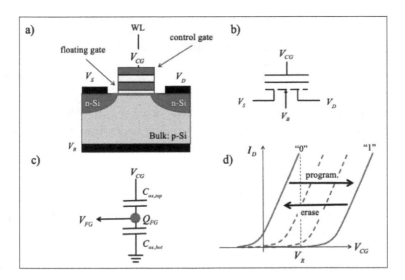

Fig. 10.6 Illustration of a floating gate MOSFET of the type used in a nonvolatile, Flash memory. Upper left a): The 1 transistor memory cell with two gates, one of them floating. Upper right b): The circuit schematic symbol for the floating gate MOSFET. Lower left c): Illustration of the gate stack showing the two gate oxide capacitors and the possibility of a stored charge, Q_{FG}, on the floating gate. Lower right d): The transfer characteristics of the floating gate MOSFET showing the programmed and unprogrammed states.

To understand how Q_{FG} affects the transistor, we can relate the voltage on the floating gate, V_{FG}, to the voltage on the control gate, V_{CG} with the circuit in Fig. 10.6c. We can solve for V_{FG} by superposition. To find how the charge, Q_{FG}, affects V_{FG}, we first assume that $V_{CG} = 0$, in which case the two capacitors are in parallel and the voltage across the two parallel capacitors is

$$V'_{FG} = \frac{Q_{FG}}{C_{ox,top} + C_{ox,bot}}. \tag{10.4}$$

To find how the voltage, V_{CG}, affects V_{FG}, we assume that $Q_{FG} = 0$, in which case, elementary circuit analysis (voltage division for two capacitors in series) gives V_{FG} in terms of V_{CG} as

$$V_{FG}'' = \frac{C_{ox,top}}{C_{ox,top} + C_{ox,bot}} V_{CG} . \tag{10.5}$$

By adding these two results, we find the relation between the floating and control gate voltages to be

$$V_{FG}' + V_{FG}'' = V_{FG} = \frac{Q_{FG}}{C_{ox,top} + C_{ox,bot}} + \frac{C_{ox,top}}{C_{ox,top} + C_{ox,bot}} V_{CG} , \tag{10.6}$$

which can solved for the control gate voltage to find

$$V_{CG} = \frac{C_{ox,top} + C_{ox,bot}}{C_{ox,top}} V_{FG} - \frac{Q_{FG}}{C_{ox,top}} . \tag{10.7}$$

To turn the transistor on, the voltage on the floating gate will have to reach a critical value, $V_{FG} = V_{TF}$; the corresponding control gate voltage is the threshold voltage of the floating gate MOSFET,

$$V_T = \frac{C_{ox,top} + C_{ox,bot}}{C_{ox,top}} V_{TF} - \frac{Q_{FG}}{C_{ox,top}} . \tag{10.8}$$

The difference in the threshold voltage with and without Q_{FG} (i.e. programmed vs. unprogrammed) is

$$\Delta V_T = -\frac{Q_{FG}}{C_{ox,top}} . \tag{10.9}$$

A negative stored charge is seen to increase the threshold voltage.

Now we can understand the operation of this memory cell. The solid lines in Fig. 10.6d are for a memory cell that stores one bit, a "0" or a "1". As shown in the figure, in the unprogrammed, "0" state with $Q_{FG} = 0$, the MOSFET has a negative threshold voltage. In the programmed, "1" state with $Q_{FG} < 0$, the threshold voltage is positive. When the control gate voltage (the wordline voltage) to V_R, then if the MOSFET is in the "0" state, a significant current flows, but if the MOSFET is in the "1" state, little or no current flows.

With modern technology, the threshold voltage can be controlled precisely. The four lines in Fig. 10.6d show how four 2 bit words, 00, 01, 10, and 11, can be stored in a single transistor that can be programmed with four different V_T's. By applying different wordline voltages, $V_{CG} = V_{R_x}$, the current can be sensed and the state of the memory cell determined. If

I_{max} is the drain current sensed when $Q_{FG} = 0$, and ΔI is the change in current that can be sensed, then the number of bits is

$$n = \log_2 \left(\frac{I_{max}}{\Delta I} + 1 \right). \tag{10.10}$$

and 2^n binary words can be stored. Currently, the most advanced memories use $n = 4$ bits, which requires that 16 threshold voltages be programmed.

Figure 10.7 shows how we program the device (i.e. put electrons on the floating gate) and how we erase the device (i.e. remove electrons from the floating gate). To program the memory cell, we ground the source, drain, and body and put a large, positive, programming voltage V_P, on the control gate (the wordline). Electrons uniformly tunnel through the thin bottom oxide (the *tunnel oxide*). This process can be precisely controlled so that specific amounts of charge, Q_{FG} can be placed on the floating gate to permit the storage of several bits of information. Typical programming times are a few hundreds of microseconds, which is much longer that the typical read times of a few tens of nanoseconds.

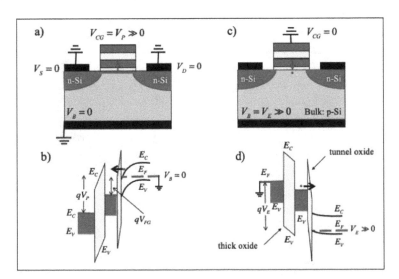

Fig. 10.7 Illustration of how a floating gate memory cell is programmed and erased. a): The programming procedure in this a large positive programming voltage, V_P, is placed on the word line (control gate). b): Energy band diagram illustrating the programming process. c): The erase procedure in which a large positive erase voltage, V_E, is placed on the body. d): Energy band diagram illustrating the erase process.

Figure 10.7b shows the energy band diagram at the beginning of the programming operation when the charge on the floating gate is negligible. (See the Lecture 10 Exercise for the energy band diagram at the end of the programming operation.) The large programming voltage ($V_P \approx 20$ V) on the control gate, pulls the energy bands down so that electrons can tunnel through the thin tunnel oxide from the channel to the floating gate. The thickness of the tunnel oxide is less than 10 nanometers. Note that the strong electric field in tunnel oxide (the steep slope of $E_c(x)$ in the tunnel oxide) reduces the thickness of the layer that electrons need to tunnel through. This type of electric-field dependent tunneling is known as *Fowler-Nordheim tunneling*. After the programming operation, the gate voltage is reduced, and electrons are confined on the floating gate by the large energy barriers from the conduction band of the polysilicon gate to that of the SiO_2. The charge on the floating gate can be retained for many years, which is what makes this a nonvolatile memory.

Figure 10.7c shows how the memory cell is erased. The control gate is grounded and a large, positive erase voltage, V_E, is applied to the p-type substrate, and the source and drain are left floating. The positive voltage on the body attracts electrons, so electrons on the floating gate tunnel into the p-type substrate. Typical erase times are a few milliseconds.

Figure 10.7d shows the energy band diagram at the beginning of the erase operation when the charge on the floating gate is substantial. The large erase voltage lowers the energy bands in the p-type Si, which produces a strong electric field in the tunnel oxide. Electrons stored on the floating gate tunnel through the thin tunnel oxide by Fowler-Nordheim tunneling to the p-type substrate.

10.7   NAND flash memory arrays

Just as ROM arrays can be organized as logical NOR or logical NAND strings, the same applies to Flash memories. NAND Flash offers the highest density, and for that reason is preferred when storage density is the top concern. Figure 10.8 shows a small section of a NAND Flash array organized like the NAND ROM array in Fig. 10.5. In contrast to the NAND ROM, there is a floating gate transistor at every memory location in the NAND Flash array.

To read the memory cell at (WL, BL) = (1, 2), all of the other wordlines would be set to a high voltage to turn them on irrespective of their

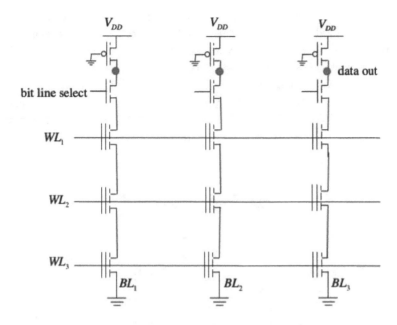

Fig. 10.8 Section of a NAND Flash array showing three bitlines and three wordlines. In contrast to the NAND ROM array shown in Fig. 10.4, there is a transistor at the intersection of every word and bit line.

programmed state. Wordline 1 would be set to the read voltage, V_R. If the cell has been programmed with a "0", the transistor will turn on, and the data out node of BL_2 will be connected to ground. If the cell has been programmed with a "1", the transistor will be off, and the output will be high. To maximize the storage density, the word and bit lines are long. The result is large RC time constants and long read times. Because of the slow response, data is not read from (or written to) one specific cell. Instead, cells along the same wordline are read in parallel. In fact, even larger groups of data are accessed in parallel — pages, which consist of several strings, or blocks, which consist of several pages.

As transistor dimensions were scaled down, the density of NAND Flash memory arrays increased to over 1 Gb/mm^2, but when critical feature sizes reached 15 nm, it became expensive to continue to shrink the transistor size. Short channel effects (i.e. the 2D electrostatics discussed in Lecture 3) also became difficult to manage. To increase NAND Flash density, 3D architectures were introduced.

In 3D NAND arrays, the bitlines are vertical, and memory density is increased primarily by increasing the number of wordline layers. Figure 10.9 is a cross-sectional sketch of one vertical NAND string in a basic 3D NAND array. To select a specific bitline, the drain and source select transistors are turned on. In contrast to the planar MOSFETs in 2D NAND Flash, there is no connection to the p-type body of the transistors. To erase the cells, a large positive voltage is applied to the source and drain and the gates are grounded. A tunneling phenomenon known as GIDL (Gate-Induced Drain Leakage [4]) then injects holes into the polysilicon body.

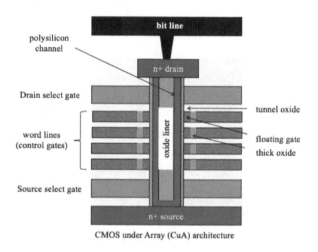

Fig. 10.9   Schematic illustration of one NAND string in a 3D NAND array. This is the cross section of a vertical, cylindrical bit line with a source at the bottom, a drain at the top, and a series of control gate / floating gate structures in between.

To produce the structure, the wordline layers are first laid down then memory holes are then etched though the layers. Next, control gate recess etch and inter polysilicon dielectric deposition is done. Polysilicon floating gates are then deposited and etched back followed by the deposition of the tunnel oxide and the polysilicon channel which connects the source and drain. The source and drain select transistors are single gate oxide MOSFETs. The number of wordline layers can be well over 100, and the excellent electrostatics of the "gate all around" structure allow the gates to be spaced more closely than in 2D NAND arrays. A 3D NAND memory consists of an array of these structures.

Figure 10.10 shows a 3D NAND memory with a "staircase structure" that brings the wordline contacts up to the planar top layer. The drain and source select transistors are labeled DSG and SSG. To save space, the row and column decoder circuitry and the bitline sense amplifiers are placed under the memory array in a CMOS Under Array (CuA) architecture. Memory density is also increased by storing $n = 3$ bits/cell (which requires $2^n = 8$ threshold voltages) or even 4 bits/cell (16 threshold voltages). The memory capacity of 3D NAND memories currently exceeds 1 Tb per chip.

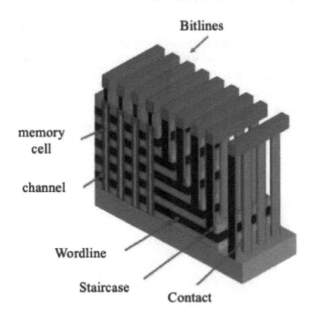

Fig. 10.10   Illustration of a 3D NAND memory array. 3D NAND technology is rapidly evolving, and many variations on this basic structure exist. (©IEEE 2020. Reprinted, with permission, from [5].)

10.8   Summary

In this lecture we have described some of the ways that transistors are used to store data. SRAMs and ROMs use conventional MOSFETs, and DRAM uses conventional MOSFETs with specially designed storage capacitors. Flash memories use a special type of MOSFET with a floating gate. The different types of memories have much different characteristics in terms of the density of information that can be stored, the length of time that it can be stored, and the time to access the data. No one memory can meet all of the needs, so information processing systems make use of a hierarchy of different types of memory. The major themes of memory technology development has been in increasing the bit density, so that more and more information can be stored.

We have described only a few important memory technologies, but there are several types of memory. Some are MOSFET-based, such as the ferro-electric MOSFET used in FeRAMs. There are non-semiconductor memories as well, such as Magnetic RAM (MRAM), Resistive RAM (ReRAM), phase change memory (PCM), in addition to magnetic disc drives and even magnetic tape. Semiconductor memory dominates, even when fairly large amounts of data must be stored. The semiconductor memory market is dominated by DRAM and NAND Flash chips. Indeed, 3D NAND memory, a *tour de force* of silicon technology, is replacing magnetic memories for many applications, and 3D DRAM is on the horizon.

Lecture 10 Exercise: Energy band diagram of a programmed Flash memory cell

Figure 10.7 showed energy band diagrams for floating gate MOSFETs being programmed and erased. In this exercise, we take a closer look at the energy band diagram of the programmed device. Figure 10.7b showed the energy band diagram at the beginning of the programming process with the control gate programming voltage, $V_{CG} = V_P$ and the voltage on the floating gate, V_{FG}. At the beginning of the programming process, there is little or no charge on the floating gate, $Q_{FG} = 0$. In this exercise, we'll make use of Eq. (10.6), which relates the control gate and floating gate voltages as

$$V_{FG} = \frac{Q_{FG}}{C_{tot}} + \left(\frac{C_{ox,top}}{C_{tot}}\right) V_{CG},$$

where

$$C_{tot} = C_{ox,top} + C_{ox,bot}.$$

For this exercise assume:

$$V_P = 20 \quad \text{V}$$
$$t_{ox,bot} = 7 \quad \text{nm}$$
$$t_{ox,top} = 14 \text{ nm}$$
$$\kappa_{ox} = 4,$$

which gives

$$C_{ox,bot} = \frac{\kappa_{ox}\epsilon_0}{t_{ox,bot}} = 5.06 \times 10^{-7} \text{ F/cm}^2$$

$$C_{ox,top} = \frac{\kappa_{ox}\epsilon_0}{t_{ox,top}} = 2.53 \times 10^{-7} \text{ F/cm}^2$$

$$C_{tot} = C_{ox,top} + C_{ox,bot} = 7.59 \times 10^{-7} \text{ F/cm}^2$$

Let's first find V_{FG} at the beginning of the programming process. From Eq. (10.6), we find

$$V_{FG} = \frac{Q_{FG}}{C_{tot}} + \left(\frac{C_{ox,top}}{C_{tot}}\right) V_{CG} = 0 + \left(\frac{2.53}{7.59}\right) 20 = 6.67 \text{ V}.$$

This is the situation shown in Fig. 10.7b.

Next, let's find V_{FG} at the end of the programming process assuming that $Q_{FG} = -q \times 10^{13}$ C/cm^2. From Eq. (10.6), we find

$$V_{FG} = \frac{(-1.6 \times 10^{19})10^{13}}{7.59 \times 10^{-7}} + \left(\frac{2.53}{7.59}\right) 20 = -2.11 + 6.67 = 4.56 \text{ V}.$$

We see that because of the negative charge on the floating gate, the energy of the floating gate has increased by 2.11 eV.

Finally, after the programming process, V_{CG} is grounded, and we find

$$V_{FG} = -2.11 \text{ V}.$$

The corresponding energy band diagram is sketched below.

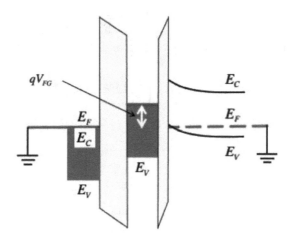

The energy band diagram of a programmed floating gate MOSFET with significant negative charge on the floating gate.

10.9   References

For an introduction to memory chip design including memory cells, sense amplifiers and drivers, and address decoders, see Chapter 16 in [1] or Chapters 8 and 9 in [2].

[1] Adel S. Sedra and Kenneth C. Smith, *Microelectronic Circuits*, 7th Ed., Oxford University Press, New York, 2015.

[2] David A. Hodges, Horace G. Jackson, and Resve Saleh, *Analysis and Design of Digital Integrated Circuits*, 3rd Ed., McGraw-Hill, Boston, 2004.

A good discussion of the history and current status of DRAMs can be found at:

[3] Wikipedia, *Dynamic Random Access Memory*, `https://en.wikipedia.org/wiki/Dynamic_random-access_memory`.

Good reviews of NAND Flash technology are:

[3] C. Compagnoni, A. Goda, A. Spinelli, P. Feeley, A. Lacaita, and A. Visconti, "Reviewing the Evolution of the NAND Flash Technology," *Proc. of the IEEE*, **105**, pp. 1609–1633, 2017.

[4] A. Goda, "3D NAND Technology Achievements and Future Scaling Perspectives," *IEEE Trans. Electron Devices*, **67**, pp. 1373–1381, 2020.

[5] N. Chandrasekaran, N. Ramaswamy, and C. Mouli, "Memory Technology: Innovations needed for continued technology scaling and enabling advanced computing systems," *IEEE Intern. Electron Devices Meeting*, pp. 10.1.1–10.1.8, 2020.

For a discussion of GIDL, the physical mechanism used to erase a NAND pillar, see Sec. 10.4.4 in:

[6] J.A. del Alamo, *Integrated Microelectronic Devices*, Pearson, New York, 2018.

Lecture 11

Heterostructure Transistors

11.1   Introduction

The focus in these lectures has been on Si MOSFETs, primarily for applications in digital electronics, but transistors also play a critical role in high-frequency wireless communications. For such applications, different types of transistors made with different semiconductors out-perform Si MOSFETs. Two important types of transistors are the HEMT and the HBT. The High Electron Mobility Transistor (HEMT) is a field-effect transistor that operates much like the MOSFETs we have been discussing. The Heterojunction Bipolar Transistor (HBT) is a high-performance type of Bipolar Junction Transistor (BJT). It is sometimes said that there are two different types of transistors — voltage controlled transistors (FETs) and current controlled transistors (HBTs and BJTs). Actually, both are barrier controlled devices; in the first, the barrier is controlled by a voltage applied to the gate contact, and in the second, the barrier is controlled by a current injected into the base [1].

This lecture is a very brief look at two important barrier controlled transistors: the HEMT and the HBT. Both the HEMT and the HBT are *heterostructure devices* by which we mean that a single device is composed of different semiconductor materials. These devices were enabled by the development of sophisticated epitaxial growth technologies in the 1970s and 1980s. The treatment in this lecture is not nearly as complete as our discussion of MOSFETs — the goal is only to make you aware that these devices exist and of the fact that they operate by controlling an energy barrier, just as MOSFETs do. We begin with the HEMT.

11.2 Modulation doping

Before discussing the HEMT, let's discuss the innovation that led to its invention. To make a field-effect transistor, we need carriers in the channel. Figure 11.1 shows two different ways of accomplishing this. Figure 11.1a is a MOSFET; as we have discussed, a positive gate voltage lowers the energy in the channel, which allows electrons to flow in from the source. Silicon is not the best semiconductor for very high-frequency applications — III-V semiconductors have much lighter electron effective masses, so the mobilities and electron velocities are significantly higher. Until recently, however, it was not possible to make MOS structures with III-V semiconductors because of defects at the insulator-semiconductor interface, so a different type of field-effect transistor was needed.

The MEtal-Semiconductor Field-Effect Transistor (MESFET) shown in Fig. 11.1b is a field-effect transistor without a gate insulator. The channel is doped n-type to produce carriers at $V_G = 0$ V. The metal gate is deposited directly on the semiconductor producing a rectifying metal-semiconductor (MS) junction. A negative gate voltage reverse biases the MS junction, and the resulting energy barrier depletes the doped channel. These types of MESFETs are *depletion mode (normally on)* transistors — I_D is modulated by depleting the semiconductor channel. *Enhancement mode (normally off)* MESFETs are also possible if the built-in potential at $V_G = 0$ fully depletes the channel, so that a small forward bias on the gate (too small to turn on the MS diode), undepletes the channel and allows current to flow.

In the 1970s, GaAs MESFETs became important high-frequency devices because of the light effective mass and higher mobility of GaAs in comparison to Si. The full capability of the material could not, however, be exploited because the channel had to be doped, and doping introduces ionized impurities that scatter electrons and reduces their mobility. For example, in undoped GaAs, $\mu_n \approx 8500 \text{ cm}^2/\text{Vs}$, but for $N_D = 10^{18} \text{ cm}^{-3}$, $\mu_n \approx 2800 \text{ cm}^2/\text{Vs}$, about one-third of its value in undoped GaAs. *Modulation doping* is a way to introduce electrons into a semiconductor without doping it.

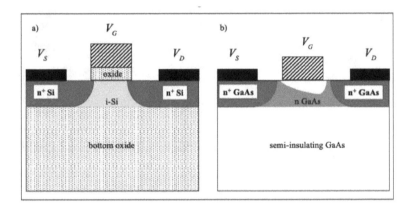

Fig. 11.1   Two n-channel FETs. Left a): A Silicon-on-Insulator (SOI) MOSFET in which a high quality gate insulator makes it possible to transfer electrons from the source and drain into an undoped channel. Right b): The GaAs MESFET, in which a doped channel is depleted by a reverse biased metal-semiconductor diode.

Figure 11.2 illustrates modulation doping. Shown in the center of Fig. 11.2a is a wide band gap semiconductor that can be doped. On the left, the Fermi level of the metal gate is indicated, and on the right is a smaller band gap, undoped semiconductor. When these three materials are put in intimate contact, electrons transfer out of the doped semiconductor where the Fermi level is the highest to the metal and to the smaller band gap, undoped semiconductor where the Fermi levels are lower. The equilibrium energy band diagram is shown on the right in Fig. 11.2b. The left half of the wide band gap doped layer is depleted because electrons transferred to the metal, and the right half is depleted because electrons transferred to the lower band gap semiconductor. Electrons have been put into the undoped semiconductor without doping it. If the structure is designed properly, the doped layer is fully depleted, so there are no electrons in this low mobility layer to conduct current. There can be, however, a high density of electrons in the high-mobility, small band gap semiconductor. These electrons provide the mobile charge, Q_n, for a FET.

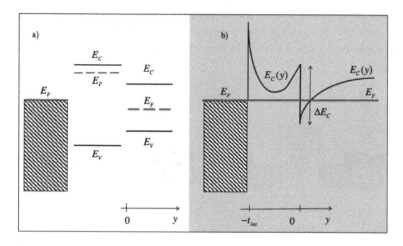

Fig. 11.2   Modulation doping. Left a): Energy band diagrams for three separate layers, metal, a wide band gap doped semiconductor, and a small bandgap undoped semiconductor. Right b): The equilibrium energy band diagram for the three-layer structure after electrons have transferred to produce a constant Fermi level in the three materials. Only the conduction band is shown in Fig. 11.2b. The y-direction is normal to the channel, and the x-direction along the channel is into the page.

   Very high electron mobilities can be obtained with modulation doping. Modern epitaxial growth techniques produce wide band gap/narrow band gap interfaces in materials systems as AlGaAs/GaAs or InAlAs/InGaAs that are atomically flat — without the surface roughness of the SiO_2/Si interface, which lowers the carrier mobility in Si MOSFETs. The dopants that produce the carriers are in the wide band gap layer, not in the small band gap layer where the electrons are. They can only weakly scatter electrons in the small band gap layer. If the temperature is lowered so that phonon scattering is suppressed, then extraordinarily high mobilities can be achieved. Modulation doping was discovered in 1978 [2], and it was quickly realized that modulation doping provided a new way to make high frequency transistors [3]. Within about 10 years, electron mobilities of 10^7 cm^2/V s were being reported [4].

11.3   High Electron Mobility Transistors (HEMTs)

Figure 11.3 is a cross-section of a HEMT with a maximum frequency of operation greater than 1 THz [5, 6]. There are similarities to a MOSFET, but there is no gate dielectric in this device.

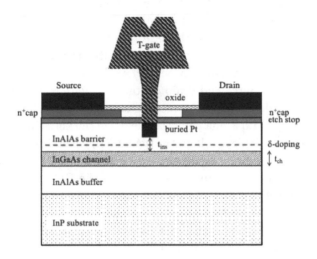

Fig. 11.3   Cross section of a III-V HEMT. The small band gap channel layer is surrounded by larger band gap layers that confine electrons to the channel. The wide band gap layer on top of the channel is delta-doped to transfer electrons to the channel. (After [6].)

The HEMT consists of layers of different III-V semiconductors grown by molecular beam epitaxy (MBE) on an InP substrate. An $In_{0.52}Al_{0.48}As$ buffer layer with $E_g \approx 1.45$ eV is first grown on the InP substrate. The buffer layer is lattice-matched to InP to minimize the density of crystal defects. Next is a 10 nm thick, In-rich InGaAs channel, which has a very small band gap and a very high electron mobility of $\mu_n > 10{,}000$ cm^2/Vs. The channel is *pseudomorphic*, which means that it is not lattice-matched to the $In_{0.52}Al_{0.48}As$ buffer, but it is thin enough that the lattice can distort to match the lattice of the buffer layer without there being sufficient strain to generate lattice defects. It is important that the channel layer be thin to achieve strong electrostatic control by the gate.

On top of the channel, an $In_{0.52}Al_{0.48}As$ barrier layer is grown by MBE. This is the wide band gap layer used for modulation doping. Rather than doping this layer uniformly, it is *delta-doped* with a plane of Si atoms, which allows a higher concentration of electrons to be transferred to the channel and increases the breakdown voltage of the MS junction. On top of the $In_{0.52}Al_{0.48}As$ barrier is an InP etch stop layer, which is used for the selective chemical etches needed for the gate recess etch, and on top of the etch stop layer is a heavily doped, n^+ cap layer to facilitate low resistance ohmic contacts to the source and drain.

Device fabrication begins with *mesa isolation,* a chemical etch through all of the layers to InP substrate, which isolates the active region of one transistor from another. Next, a Ni/Ge/Au layer is evaporated and patterned to form the source and drain contacts. A gate recess etch thins the barrier layer in order to place the metal gate close to the channel. The distance, t_{ins}, in Fig. 11.3 is analogous to the oxide thickness in a MOSFET; it needs to be thin to achieve a high transconductance and to minimize short channel effects. The large *T-gate* on top of the actual Pt gate reduces the gate resistance. Gate resistance has no role in the DC *IV* characteristics, but a low gate resistance is critical for high-frequency performance. The stem of the T-gate is tall to reduce the parasitic gate to drain capacitance.

The wide band gap layer on top of the channel is analogous to the oxide of a MOSFET, but its band gap and band offset, ΔE_c, (see Fig. 11.2) are much smaller. Because there is no gate insulator, only small positive voltages can be applied to the gate, otherwise too much gate current flows. Because ΔE_c is rather small, under high V_{DS}, electrons in the channel can gain energy and scatter to the barrier layer. Nevertheless, the HEMT is a barrier controlled transistor with *IV* characteristics look much like those of a MOSFET's, as shown in Fig. 11.4. The virtual source model developed for Si MOSFETs does an excellent job of fitting the *IV* characteristics of this HEMT. The fitted apparent mobility is, however, only 220 cm^2/Vs, far below the scattering limited mobility of the InGaAs channel, $\mu_n >$ 10,000 cm^2/Vs, but because of the small electron effective mass, the fitted injection velocity, $v_{inj} = 3.5 \times 10^7$ cm/s, is much higher than for a Si MOSFET. See Lecture 19 in [7] for a virtual source analysis of the *IV* data shown in Fig. 11.4.

Several different materials systems are used for HEMTs. The first laboratory demonstration and commercial products used AlGaAs/GaAs for the wide/narrow band gap layers. Today, a widely-used material system is the InAlAs/InGaAs/InP system shown in Fig. 11.3. More recently, HEMTs have been developed in the AlGaN/GaN system, which is an interesting

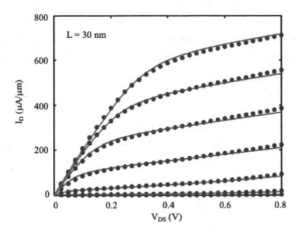

Fig. 11.4 *IV* characteristics of an $L = 30$ nm III-V HEMT. The points are the measured data [5], and the lines are the MVS analysis and plot were provided by Dr. S. Rakheja, MIT, 2014. Data provided by D.-H. Kim. Used with permission. See [7] for a VS analysis of this data.

system because the piezoelectric polarization at the AlGaN/GaN interface creates the mobile charge in the channel, so there is no need to dope the wide band gap AlGaN layer.

HEMTs have applications in smart phones, cellular base stations, fiber-optic communications systems, wireless local area networks, satellite communications, radar, radioastronomy, sensing, and defense electronics [8, 9]. The band gaps of AlGaN and GaN are large, and the electron velocities are high making AlGaN/GaN HEMTs well-suited for high-power/high-voltage RF applications as well as for power electronics. The total number of HEMTs manufactured is small compared to the number of Si MOSFETs. but by addressing applications for which Si MOSFETs are unsuitable or non-optimum, they play a critical role in modern electronics.

11.4   PN heterojunction diodes

Figure 11.5 is a cross-section of a silicon double diffused bipolar junction transistor (BJT), the first transistor technology used in integrated circuits. While this device is not a heterostructure transistor, it illustrates the basic components of any bipolar transistor The three terminals are: 1) the emitter (E), which is analogous to the source of a MOSFET, 2) the base (B), which is analogous to the channel, and 3) the collector (C), which is analogous to the drain. The heart of the transistor is the vertical npn structure, which consists of two pn junctions, one between the emitter and base, and one between the base and collector. The p-type region is shared between the two pn junctions. This npn BJT is analogous to an n-channel MOSFET. There is a complementary pnp BJT analogous to a p-channel MOSFET. Just as we had to understand modulation doping before we could understand HEMTs, we need to understand pn junctions, before we can understand bipolar transistors. Our discussion will be brief; see any semiconductor textbook for a thorough discussion of pn junctions (e.g. [10]).

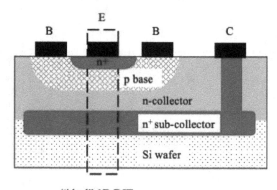

"ideal" 1D BJT

Fig. 11.5 Cross-section of a double diffused Si bipolar junction transistor with an n-type emitter (E), p-type base (B), and n-type collector (C). The intrinsic device is the vertical npn structure shown in the dashed line box. The notation, n^+ means that the n-type doping density is high.

Figure 11.6 shows the energy band diagrams for homo and hetero pn junctions. On the left in Figs. 11.6a and 11.6b are the individual semiconductors. Note that in Fig. 11.6b the n-type semiconductor has a larger band gap than the p-type semiconductor. The two conduction bands are shown to line up; this is the ideal case for an npn heterojunction BJT (HBT), but band offsets are material-specific. In the n-type semiconductors, the equilibrium electron concentration is $n_0 = N_D$ and the minority hole concentration is $p_0 = n_{in}^2/N_D$. In the p-type semiconductors, the equilibrium hole concentration is $p_0 = N_A$ and the minority electron concentration is $n_0 = n_{ip}^2/N_A$. As shown in Figs. 11.6c and 11.6d, when we conceptually

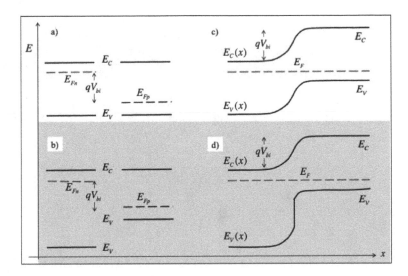

Fig. 11.6 Equilibrium energy band diagrams for pn junctions. Left a) and b): Separate n-type and p-type semiconductors. In a) both semiconductors are the same, but in b) the n-type semiconductor has a wider band gap. Right c) and d): The resulting equilibrium energy band diagrams when the two semiconductors are conceptually joined.

bring the two semiconductors together, electrons transfer from the n-type semiconductor with the higher E_F to the p-type semiconductor with the lower E_F, and holes transfer from the p-side to the n-side. The charge transfer lowers the electrostatic potential of the n-type semiconductor until its Fermi level aligns with that of the p-side. Note from Figs. 11.6c and 11.6d that electrons see an energy barrier that prevents them from diffusing to the p-side, and holes see an energy barrier that keeps them on the p-side (hole energy increases downward on this electron energy plot). The *built-in potential* difference between the two sides of the junction is

$$qV_{bi} = E_{Fn} - E_{Fp}. \tag{11.1}$$

Next, let's examine the diode current as a function of the voltage applied to the contacts.

Figure 11.7 shows the Np diode under forward bias — a positive voltage, V_A, has been applied to the p-side of the junction. (The capital N is used to identify the wide band gap side of the junction.) As shown by the energy band diagram in Fig. 11.7a, the positive voltage lowers the hole quasi-Fermi level, F_p, on the p-side and the conduction and valence bands follow. The regions away from the junction are very conductive because on the n-side the electron concentration is high and on the p-side the hole concentration is high. The potential in the semiconductor changes in the most resistive part of the diode — the transition region around the junction. The result is that the energy barrier is lowered by qV_A. Recall that the probability that an electron can surmount an energy barrier of height, E_b, is $\exp(-E_b/k_BT)$. The barrier has been lowered by qV_A, so the probability of electrons hopping over the barrier is increased by a factor of $\exp(qV_A/k_BT)$. We expect, therefore, that the minority electron concentration at the beginning of the neutral p-region will be increased by a factor of $\exp(qV_A/k_BT)$.

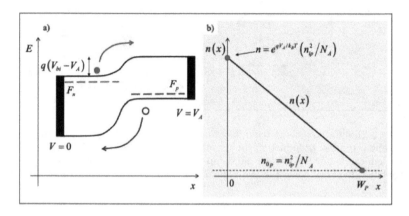

Fig. 11.7   A forward-biased Np heterojunction diode. Left a): Energy band diagram. Right b): The minority electron concentration vs. position in the neutral part of the p-region — away from the transition region.

Figure 11.7b shows $n(x)$, the minority electron concentration vs. position on the neutral p-side of the diode. We assume that the ohmic contact to the p-region maintains equilibrium, so $n(x = W_p) = n_{0p} = n_{ip}^2/N_A$. Electrons are injected over the barrier to the beginning of the p-type region, and then they diffuse across the neutral p-region. For a good transistor, the p-region is short, so very few electrons recombine with holes as they diffuse across. The electron current at the beginning of the p-region must be identical to the electron current at $x = W_B$; the linear profile shown for $n(x)$ ensures that this is the case.

Beginning with Fick's Law of diffusion, Eq. (7.3), we find the electron current as

$$I_n = qA_D\left[-D_n dn(x)/dx\right]$$
$$= qA_D\frac{D_n n_{ip}^2}{W_P N_A}\left(e^{qV_A/k_BT} - 1\right), \tag{11.2}$$

where A_D is the area of the diode. The energy barrier for holes is also reduced, so there is an analogous hole current of

$$I_p = qA_D \left[-D_p dp(x)/dx\right]$$

$$= qA_D \frac{D_p n_{in}^2}{W_N N_D} \left(e^{qV_A/k_B T} - 1\right),$$

(11.3)

and the total diode current is

$$I_D = I_n + I_p = qA_D \left[\frac{D_n n_{ip}^2}{W_P N_A} + \frac{D_p n_{in}^2}{W_N N_D}\right] \left(e^{qV_A/k_B T} - 1\right).$$

(11.4)

The physical picture for the electron contribution to the steady-state diode current is as follows. Electrons are injected from the n-region over the barrier to the p-type region. They diffuse across the p-region, leave through the contact, travel around the external circuit and re-enter the n-type region where they replace the electrons that were injected over the barrier.

We have been careful to distinguish between the intrinsic carrier concentrations on the two sides of the junction. Recall that $n_i^2 \propto \exp(-E_g/k_B T)$. Assuming that the effective densities-of-states for the two semiconductors are similar, then

$$n_{in}^2/n_{ip}^2 \approx e^{-\Delta E_g/k_B T},$$

(11.5)

where $\Delta E_g = E_{gn} - E_{gp}$. For bipolar transistors, it will be important to know the relative magnitudes of I_n and I_p. From Eqs. (11.2) and (11.3), we find

$$\frac{I_n}{I_p} = \frac{D_n W_N}{D_p W_P} \times \frac{N_D}{N_A} \times \frac{n_{ip}^2}{n_{in}^2}.$$

(11.6)

To use this diode as the emitter-base junction of a BJT, we need I_n/I_p to be at least 10, preferably much higher. For most semiconductors, $D_n > D_p$, and the base is thin in a BJT, so $W_P < W_N$ and the first factor on the RHS of Eq. (11.6) is greater than one, but it is not large enough to make I_n/I_p large. If this is a homojunction diode, then $n_{ip} = n_{in}$ and the last factor on the RHS of Eq. (11.6) is one. We conclude that to make I_n/I_p large, we require $N_D \gg N_A$. In practice, the n-region is doped as heavily as possible, but the base cannot be doped too heavily if we are to keep I_n/I_p large.

Next, consider a heterojunction diode in which the n-region band gap is only 0.3 eV larger than the p-region band gap. That is 11.5 $k_B T$, so

$$n_{ip}^2/n_{in}^2 = e^{11.5} = 10^5,$$

(11.7)

which essentially means that we can dope the base as heavily as we wish, and I_n/I_p will still be large. Doping the base heavily lowers its resistance. As shown in Fig. 11.5, the current in the p-layer has to flow laterally up to the contacts on the surface, and minimizing these lateral volt drops improves the performance of the transistor.

We've discussed the forward-biased pn junction, which serves as the emitter-base junction of a BJT; next, let's discuss the reverse-biased pn junction shown in Fig. 11.8. A reverse bias, V_R, a positive voltage on the n-side, increases the height of the energy barrier so that it is essentially impossible for any electrons from the n-side to surmount the barrier. There are a few minority carrier electrons on the p-side, however. The number is small, $n_{0p} = n_{ip}^2/N_A$ away from the junction. Electrons on the p-side near the top of the energy barrier can fall down and exit through the contact to the n-side. We see that a reverse biased pn junction "collects" minority carriers. The density of electrons on the p-side is very small, so the result is only a small reverse biased leakage current. In a BJT, however, there can be a high density of electrons in the p-region, and the reversed-biased collector-base junction can collect them.

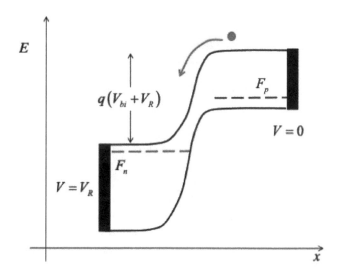

Fig. 11.8   A reverse-biased Np heterojunction diode. This is the same pn junction as in Fig. 11.7, but now with the p-side grounded and a positive voltage, V_R applied to the n-side.

11.5   Heterojunction Bipolar Transistors (HBTs)

The *IV* characteristics of an NpN HBT are sketched in Fig. 11.9. The output characteristics (Fig. 11.9a) are similar to those of a MOSFET, but the voltage at which I_C saturates is only a few $k_B T/q$, which is much smaller than the V_{DSAT} of a MOSFET. Note also that the family of $I_C(V_{CE})$ characteristics is produced by linearly increasing the <u>base current</u>, not by linearly increasing V_{BE}. The transfer characteristics (Fig. 11.9b) are also similar to those of a MOSFET, but the base current, $I_B(V_{BE})$, is also plotted; I_B is typically 10–100 times smaller than I_C, but in contrast to the gate current of a MOSFET, it cannot be ignored. Some roll-off in $I_C(V_{BE})$ is observed at high V_{BE}, but it is much smaller than the roll-off of I_D at high V_{GS}. For the HBT, the roll-off is mostly due to R_E, the emitter series resistance, which can be minimized by process design, but for a MOSFET, the drain current roll-off above threshold is intrinsic to the device. For an HBT, I_C varies exponentially with input voltage, but for a MOSFET, I_D only varies exponentially when the input voltage is below V_T.

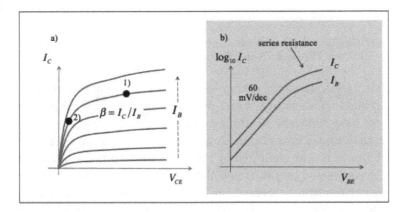

Fig. 11.9   *IV* characteristics of a bipolar transistor. Left a): Output characteristics — I_C vs. collector-emitter voltage. Right b): Transfer characteristics, $\log_{10} I_C$ vs. base-emitter voltage. Also shown is $I_B(V_{BE})$. For the output characteristics in a), two operating points are identified: 1) in the active region and 2) in the saturation region.

Figure 11.10 is an equilibrium energy band diagram of an NpN HBT. Note the similarity to the equilibrium energy band diagram of a MOSFET as shown in Fig. 3.3. The main difference is how the energy barrier is modulated. Next, we will examine the energy band diagrams at points 1) and 2) on the output characteristics sketched in Fig. 11.9a.

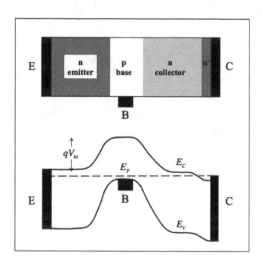

Fig. 11.10   Equilibrium energy band diagram for A 1D, NpN HBT. Top: The physical structure. Bottom: The energy band diagram.

Figure 11.11a shows the energy band diagram for point 1) in Fig. 11.9a. The E-B junction is forward-biased so that the emitter injects electrons into the p-type base, but these electrons don't leave through the p-type base contact, as they would in a pn junction. The base is thin, and the electrons which diffuse across are collected by the reverse-biased B-C junction. Recall that a reverse-biased pn junction collects minority carriers. In equilibrium, the minority electron concentration in the p-type base is extremely small, but in this case, the forward-biased E-B junction increases their concentration greatly. The E-B junction injects electrons into the base, the collector collects them, and this is the collector current, I_C. Note the similarity of the HBT energy band diagram for large V_{CE} to that of a MOSFET for a large V_{DS} (Fig. 7.5). For point 1) in Fig. 11.9a, the collector current is the electron component of the forward biased E-B junction. Analogous to Eq. (11.2), we find:

$$I_C = qA_E \frac{D_n n_{iB}^2}{W_B N_{AB}} e^{qV_{BE}/k_B T} , \qquad (11.8)$$

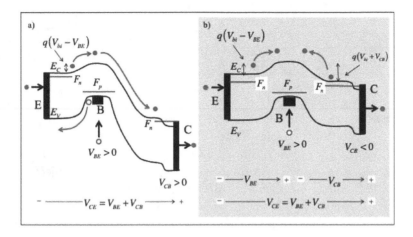

Fig. 11.11 Energy band diagrams for an HBT biased two different ways. Left a): In the active region where the emitter-base junction is forward biased and the base-collector junction is reverse biased. This bias corresponds to point 1) in Fig. 11.9a. Right b): In the saturation region where both the emitter-base and base-collector junctions are forward biased. This bias corresponds to point 2) in Fig. 11.9a.

where A_E is the area of the E-B junction, n_{iB}^2 is the intrinsic carrier concentration in the base, W_B is the thickness of the neutral part of the base, and V_{BE} is forward bias on the intrinsic B-E junction, which is assumed to be greater than a few $k_B T/q$ so that the -1 in Eq. (11.2) can be dropped. For high currents, $V_{BE} = V'_{BE} - I_E R_E$, where R_E is the emitter series resistance, and V'_{BE} is the extrinsic base-emitter voltage. (The base current is small, so we neglect the voltage drop in the base series resistance.) Equation (11.8) is analogous to $I_D = W C_g v_{inj}(V_{GS} - V_T)$ for a MOSFET. Note the important point that for a BJT the collector current is an exponential function of the input voltage, V_{BE}, while for a MOSFET above threshold, the drain current is a linear function of the input voltage, V_{GS}.

For a MOSFET, the DC gate current is very small, but for a BJT, the base current must be considered. There are two current components in the forward-biased E-B junction: the electron current injected into the base, $I_n \approx I_C$, and the hole current, I_p, injected into the emitter. When a hole is injected from the base to the emitter, it must be replaced by a hole that enters from the base terminal (or equivalently, an electron in the valence band of the base must leave through the base contact to create a hole in the base). We conclude that $I_p = I_B$, so

$$\beta \equiv \frac{I_C}{I_B} = \frac{I_n}{I_p} = \frac{D_n W_E}{D_p W_B} \times \frac{N_{DE}}{N_{AB}} \times \frac{n_{iB}^2}{n_{iE}^2}, \tag{11.9}$$

where we have used Eq. (11.6). A current gain, β, of at least 10 is generally required. As was shown by Eq. (11.7), an emitter with a band gap only a few tenths of an eV greater than the base can make β very large — even if $N_{AB} \gg N_{DE}$, which is needed for the low base resistance necessary for good high frequency performance. In practice, the back-injected hole current, I_p, can be negligible in an HBT, but this does not mean that $\beta \to \infty$ because there is a small chance that electrons diffusing across the base will recombine with a hole. When that happens, the hole is replaced by one the enters through the base contact. In practice, base recombination limits β in an HBT.

Bias condition 1) in Fig. 11.9a is called the *forward active region* and is analogous to the saturation region of a MOSFET. Bias condition 2) is known as the *saturation region of a BJT*. The energy band diagram is shown in Fig. 11.11b. The voltages across the two junctions control the performance of an HBT. In the common emitter configuration we are examining, the input voltage, V_{BE}, is across the E-B junction, but the output voltage, V_{CE}, is across two junctions. From Kirchoff's Voltage Law, we find the voltage across the collector-base junction to be $V_{CB} = V_{CE} - V_{BE}$. To turn on the E-B junction, V_{BE} is typically several tenths of a volt. When V_{CE} is less than several tenths of a volt, then $V_{CB} < 0$ and the B-C junction becomes forward biased. As shown in Fig. 11.11b, when both junctions are forward biased, electrons are injected into the base from both the emitter and the collector, and I_C is the difference in the two injected currents. As V_{CE} decreases, the forward bias on the B-C junction increases, and the net electron current, I_C, decreases as shown in Fig. 11.9a for small V_{CE}. It is important to remember that the "saturation" region for a FET occurs for large V_{DS} where I_D varies slowly with V_{DS}, and the "saturation" region for a BJT occurs for a small V_{CE} where I_C varies rapidly with V_{CE}.

In the family of HBT *IV* characteristics shown in Fig. 11.9a, the base current, I_B, is the parameter rather than the base-emitter voltage, V_{BE}, which is actually what lowers the E-B barrier and increases I_C. The reason is that I_C depends exponentially on V_{BE} (Eq. (11.8)). In a circuit, it would be very difficult to control I_C with V_{BE} because very small changes in V_{BE} would produce very large changes in I_C. From Eq. (11.8), we see that

$V_{BE} \propto \log(I_C) \propto \log(I_B)$, so it is much easier to accurately control V_{BE} by controlling I_B. Unfortunately, this has led to the confusion that MOSFETs and BJTs operate on two different principles. The operating principle of these two transistors is, however, the same [1].

Figure 11.12 is a cross-section of a representative InGaAs/InP HBT [11]. The structure is quite different from a FET, but both HEMTs and HBTs use sophisticated epitaxy to grow the multiple layers of semiconductors needed. The key device design challenges for HBTs are the need for aggressive downscaling of the emitter and collector areas to lower capacitances and the need for extremely low emitter, base, and collector resistances [12].

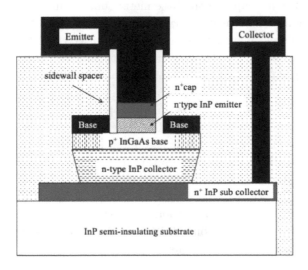

Fig. 11.12   Cross section of a III-V HBT. The small band gap base layer is surrounded by larger band gap layers for the emitter and collector. (After Fig. 1 in [11].)

For the HBT shown in Fig. 11.12, the substrate is semi-insulating InP. The heavily doped, InP n^+ sub-collector keeps the collector series resistance low. The active collector is a lightly doped n-type layer of InP; its wide band gap increases the breakdown voltage as compared to InGaAs HEMTs and SiGe HBTs. The base is a lattice-matched, small band gap heavily doped, p^+-InGaAs layer. It is doped as heavily as possible (near 10^{20} cm^{-3}). The heavy doping provides a low base resistance but results in a low carrier lifetime, which reduces β. To lower the probability that electrons recombine in the base, it is graded to produce a field that sweeps electrons across the base before they can recombine. On top of the base, a wide band gap, n-type InP emitter is grown and on top of that is a heavily doped cap layer to facilitate ohmic contact. Because the InGaAs/InP conduction bands don't line up properly, thin compositionally graded layers between the emitter and base and between the base and collector are needed. (Alternatively, GaAsSb, which has more favorable band line-ups can be used.)

The critical alignment for an HBT process is between the emitter and base contacts. The self-aligned mesa process for this device keeps the spacing between the outer edge of the emitter contact and the inner edge of the base contact small to achieve low base resistance, but large enough to minimize E-B shorts [11]. The contact resistances between the emitter metal contact and the n^+ cap and the base metal contact and the p^+ base must be exceptionally low to achieve low emitter and base resistances.

Several different material systems are used for HBTs [13]. SiGe technology is widely used in applications such as automotive radar because of its compatibility with Si CMOS technology (so-called BiCMOS technology). The InGaAs/InP material system is a well developed technology for higher frequency applications. HBTs see many applications in wireless communications, both in smartphones and in wireless base stations. They are also extensively used in fiber optic communications systems. Other applications are in wideband, high resolution analog-digital and digital-analog converters, and in monolithic millimeter-wave integrated circuits (MMICS) for applications such as radar. SiGe BiCMOS, III-V HEMTs. and III-V HBTs all have important applications in modern electronics. Each device has advantages and limitations, so the choice of technology is driven by system performance considerations.

11.6 Discussion

The high-frequency performance of both HEMTs and HBTs exceeds that of Si MOSFETs; why are both devices needed? The transconductance is the key figure of merit for a transistor used as an amplifier. For amplifying circuits, FETs are biased in the saturation region and HBTs in the active region where

$$I_D = W C_g v_{inj} \left(V_{GS} - V_T \right)$$
$$I_C = A_E \frac{n_{ib}^2}{N_{AB}} \frac{D_n}{W_B} e^{q V_{BE}/k_B T} \,.$$

The corresponding transconductances are

$$g_m^{FET} = \frac{\partial I_D}{\partial V_{GS}} = \frac{I_D}{(V_{GS} - V_T)}$$

$$g_m^{BJT} = \frac{\partial I_C}{\partial V_{BE}} = \frac{I_C}{k_B T/q}.$$

If the transistors are operated at the same current, we conclude that

$$\frac{g_m^{BJT}}{g_m^{FET}} = \frac{(V_{GS} - V_T)}{k_B T/q}. \tag{11.10}$$

Typically, V_{GS} is about 1 V, and V_T is a few tenths of a volt while $k_B T/q = 0.026$ V. We conclude that the transconductance of a bipolar transistor is more than a factor of 10 larger than that of a FET. Bipolar transistors are inherently better amplifying devices than FETs, but they consume more power and are more complicated to fabricate, so they are used where high current drive is needed.

HEMTs and HBTs are used for the high-frequency RF applications that can't be addressed with Si MOSFETs. According the Eq. (2.3), the maximum frequency of operation for a transistor is

$$\omega_T = 2\pi f_T = \frac{g_m}{C_{tot}}.$$

(The frequency, f_T, is the maximum frequency at which there is short circuit current gain. Another high-frequency metric is f_{max}, the maximum frequency of oscillation. Both are important figures of merit for circuit performance.) You may wonder why the *transit time*, the time it takes for a carrier to cross the channel in a FET or the base and collector in a BJT, does not enter in. It does in principle, but recall that for digital circuits, the speed is a few GHz and is dominated by the time it takes to charge and discharge the circuit capacitances. The DC on-current controls the speed of a digital circuit, but for RF transistors that may operate at hundreds of GHz, the transit time is important.

For a bipolar transistor, we can write the delay as

$$\begin{aligned}
\tau = \frac{1}{\omega_T} &= \frac{(C_{je} + C_{cb})}{g_m} + (\tau_b + \tau_c) + (R_E + R_C)\, C_{cb} \\
&= \left(\frac{k_B T}{q}\right) \frac{(C_{je} + C_{cb})}{I_C} + (\tau_b + \tau_c) + (R_E + R_C)\, C_{cb},
\end{aligned} \tag{11.11}$$

where C_{je} is the capacitance of the E-B junction, C_{cb} is the capacitance of the C-B junction, R_E and R_C are the emitter and collector series resistances, τ_b is the base transit time, and τ_c is the collector transit time.

Equation (11.11) helps us understand several HBT design issues. To minimize the first term, very high currents are needed; the resulting high power dissipation makes thermal effects critically important. The junction area must be very small to minimize the capacitances. The base must be thin to minimize the base transit time, and it must be heavily doped to minimize the base series resistance; R_B does not affect f_T, but it does have a strong effect on f_{max}. For high-frequency HBTs, base thicknesses of less than 30 nm are used, which makes quasi-ballistic transport important. The collector must be thin too, to minimize the collector delay but the trade-off is that this increases the collector capacitance. Careful attention to these issues has resulted in HBTs with maximum operating frequencies in excess of 1 THz [11].

11.7 Summary

This lecture has been a short introduction to two important types of transistors. The HEMT is a field-effect transistor that operates much like a Si MOSFET, but the material properties lead to excellent high-frequency performance with numerous applications in wireless and optical communication systems. The wide band gap of AlGaN/GaN HEMTs makes them well-suited for high-voltage power electronics. A key difference between a HEMT and a MOSFET is the fact that HEMTs do not employ a gate insulator, so care must be taken not to bias the gate so that the metal-semiconductor junction is forward-biased.

The bipolar junction transistor is also a barrier-controlled device. While the energy band diagram and operating principles are similar to FETs, the difference is that the barrier height is directly controlled by the voltage applied to the emitter-base junction, rather than indirectly controlled by a gate that is not in direct contact with the semiconductor, as in a FET. Modern bipolar transistors are heterostructure devices, HBTs, with a wide band gap emitter and often with a wide band gap collector too. The key difference between a FET and an HBT is the much higher transconductance of an HBT, which makes them well-suited for power amplifiers where high current drive is necessary. The "threshold voltage" of an HBT is the turn-on voltage of the E-B junction, which is set by the band gap of the base. As a result, the turn-on voltage is very well-controlled in comparison to a FET, which makes HBTs well-suited for precision applications such as A/D and D/A converters.

Lecture 11 Exercise: Quasi-ballistic transport in a BJT

The collector current in a bipolar transistor is due to electrons injected from the emitter diffusing across the base where they are collected by the collector and is given in the active region by Eq. (11.8) as

$$I_C = qA_E \left[\frac{n_{iB}^2}{N_{AB}} e^{qV_{BE}/k_BT} \right] \frac{D_n}{W_B}.$$

In equilibrium, the electron concentration at the beginning of the base is $n(0) = n_{ib}^2/N_{AB}$. Under forward bias, the energy barrier between the emitter and the base is lowered, and $n(0)$ increases exponentially to $n(0) = (n_{ib}^2/N_{AB}) \exp(qV_{BE}/k_BT)$, so we see that the term in square brackets is $n(0)$. Current is charge times velocity, so we conclude that

$$I_C = qA_E\, n(0) \langle v_x(0) \rangle\,,$$

where

$$\langle v_x(0) \rangle = \frac{D_n}{W_B}\,,$$

is the average diffusion velocity of electrons at the beginning of the base, $x = 0$. But now we see a problem. Bases can be made exceptionally thin, so it is possible that $D_n/W_B > v_T$, but electrons cannot diffuse faster than the thermal velocity because diffusion is due to random thermal motion. Equation (11.8) over-predicts the current when the base is thin. The point of this exercise is to develop a collector current equation that behaves properly in the ballistic limit.

The figure below is similar to Fig. 7.2; it shows the base of a bipolar transistor.

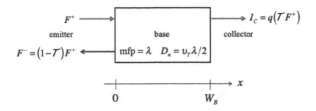

Transport across the base of a bipolar transistor biased in the active region.

The flux, F^+ is the electron flux injected from the forward based E-B junction. The fraction that transmits across produces the collector current,

$$I_C = q(\mathcal{T}F^+),$$

and the fraction that doesn't transmit across returns to the emitter,

$$F^- = (1 - \mathcal{T})\,F^+.$$

The electron concentration at the beginning of the base is

$$n(0) = \frac{(F^+ + F^-)}{v_T} = \frac{(2 - \mathcal{T})}{v_T}F^+,$$

which can be solved for F^+ to find

$$F^+ = \frac{qn(0)v_T}{(2 - \mathcal{T})},$$

from which we find the collector current as

$$I_C = q(\mathcal{T}F^+) = qn(0)\left[\left(\frac{\mathcal{T}}{2 - \mathcal{T}}\right)v_T\right]$$

$$= qn(0)\,\langle v_x(0)\rangle.$$

We have seen the factor, $[\mathcal{T}/(2 - \mathcal{T})]v_T$ before; according to Eq. (7.27) the average velocity at the beginning of the base is

$$\frac{1}{\langle v_x(0)\rangle} = \frac{1}{v_T} + \frac{1}{D_n/W_B}$$

or

$$\langle v_x(0)\rangle = \left(\frac{\mathcal{T}}{2 - \mathcal{T}}\right)v_T = \frac{D_n/W_B}{1 + (D_n/W_B)/v_T}.$$

We conclude that in the collector current equation, D_n/W_B must be replaced by $(D_n/W_B)/(1 + (D_n/W_B)/v_T)$ to find

$$\boxed{I_C = qA_E\left[\frac{n_{iB}^2}{N_{AB}}\left(e^{qV_{BE}/k_BT} - 1\right)\right]\left[\frac{D_n/W_B}{1 + (D_n/W_B)/v_T}\right].}$$

For a thick base, $D_n/W_B \ll v_T$, and we get the traditional, diffusive expression for the collector current,

$$I_C = qA_E\left[\frac{n_{iB}^2}{N_{AB}}\left(e^{qV_{BE}/k_BT} - 1\right)\right]\frac{D_n}{W_B},$$

but for a thin base, $D_n/W_B \gg v_T$, and we find the ballistic limit collector current as

$$I_C = qA_E\left[\frac{n_{iB}^2}{N_{AB}}\left(e^{qV_{BE}/k_BT} - 1\right)\right]v_T.$$

In a bipolar transistor in the active region, the diffusion of minority carrier electrons across the base is the bottleneck that controls the collector current. In a MOSFET in the saturation region, the diffusion of electrons across the low-field region near the beginning of the channel is the bottleneck for current. In both cases, the electron velocity is limited by the diffusion velocity. This discussion for bipolar transistors is very similar to the discussion in Sec. 7.6 for MOSFETs and underscores the similarity of the operating principles for MOSFETs and BJTs.

Finally, we should point out that the issue of quasi-ballistic transport first arose in the 1970s, because BJT's with very thin bases could be produced. See, for example, P. Rohr, F. A. Lindholm and K. R. Allen, "Questionability of drift-diffusion transport in the analysis of small semiconductor devices," *Solid-State Electron.*, **17**, 729, 1974. Equations like (7.27) have been used since the 1970s, to describe the velocity of electrons in the base of bipolar transistors.

11.8   References

The fact that MOSFETs and bipolar transistors operate on the same principle of controlling current by controlling the height of an energy barrier is not as widely-appreciated as it should be, but it has long been understood. See, for example:

[1] E. O. Johnson, ""The Insulated Gate Field Effect Transistor — A Bipolar Transistor in Disguise," *RCA Review*, **34**, p. 80, 1973.

The first report of modulation doping was by Dingle, et al. As the technology matured, it became possible to achieve extraordinary mobilities. Just two years after the discovery of the effect, a field-effect transistor based on modulation doping was reported by T. Mimura, et al.

[2] R. Dingle, H. L. Stormer, A. C. Gossard, and W. Wiegmann, "Electron mobilities in modulation-doped semiconductor heterojunction superlattices," *Appl. Phys. Lett.*, **33**, p. 665, 1978.

[3] T. Mimura, S. Hiyamizu, T. Fujii, and K. Nanbu, "A New Field-Effect Transistor with Selectively Doped GaAs/n $-$ Al$_x$Ga$_{1-x}$As Heterojunctions," *Jap. J. Appl. Phys.*, **19**, p. L225, 1980.

[4] L. Pfeiffer, K.W. West, H.L. Stormer, and K.W. Baldwin, "Electron mobilities exceeding 10^7 cm^2/Vs in modulation-doped GaAs," *Appl. Phys. Lett.*, **55**, p. 1888, 1989.

For a brief description of high-frequency InAlAs/InGaAs HEMTs, see:

[5] D-H Kim, J. A. del Alamo, P. Chen, W.Ha, M. Urteaga, and B. Brar, "50-nm E-mode In$_{0.7}$Ga$_{0.3}$As PHEMTs on 100-mm InP substrate with $f_{max} > 1$ THz," *Tech. Digest, International Electron Device Meeting*, p. 692, 2010.

[6] D.H. Kim and Jesus del Alamo, "30-nm InAs Pseudomorphic HEMTs on a InP Substrate With a Current-Gain Cutoff Frequency of 628 GHz," *IEEE Electron Device Letters*, **29**, no. 8, pp. 830–833, 2008.

For a VS analysis of a III-V HEMT, see Lecture 19 in:

[7] Mark Lundstrom, *Fundamentals of Nanotransistors*, World Scientific Publishing Co., Singapore, 2018.

The development of HEMTs since their invention in 1978, some of their applications, and prospects for the future, are reviewed in:

[8] Jesus A. del Alamo, "The High Electron Mobility Transistor at 30: Impressive Accomplishments and Exciting Prospects," *Proc. of the 2011 Int. Conf. on Compound Semiconductor Manufacturing Technology*, http://hdl.handle.net/1721.1/87102, 2011.

[9] Jesus A. del Alamo,, "Nanometre-scale electronics with III-V compound semiconductors," *Nature*, **479**, p. 317, 2011.

For an introduction to pn junctions, see

[10] Robert F. Pierret, *Semiconductor Device Fundamentals*, Addison-Wesley Publishing Co., 1996.

For more about HBT technologies and applications, see:

[11] M. Urteaga, Z. Griffith, M. Seo, J. Hacker, and M. Rodwell, "InP HBT Technologies for THz Integrated Circuits," *Proc. IEEE*, **105**, pp. 1051–1067, 2017.

[12] M. Rodwell, M. Urteaga, T. Mathew, D. Scott, D. Mensa Q. Lee, J. Guthrie, Y. Betser, S.C. martin, R.P. Smith, S. Jaganatrhan, S. Krishnan, S.I. Long, P. Pullela, B. Agarwal, U. Bhattacharaya, L. Samoska, and M. Dahlkstrom, "Submicron Scaling of HBT's," *IEEE Trans. Electron Dev.*, **48**, pp. 2606–2624, 2001.

[13] M. Bozanic and S. Sinha, "Emerging Transistor Technologies Capable of Terahertz Amplification: A Way to Re-Engineer Terahertz Radar Sensors," *Sensors*, **19**, p. 2454, 2019.

Lecture 12

Transistor Reliability

12.1 Introduction
12.2 Negative Bias Temperature Instability (NBTI)
12.3 Hot Carrier Degradation (HCD)
12.4 Time Dependent Dielectric Breakdown (TDDB)
12.5 Discussion
12.6 Summary
12.7 References

12.1   Introduction

Companies that manufacture integrated circuits and discrete transistors must assure their customers that these devices will function reliably for some time; 10 years is typical. This lecture is a short introduction to the physics of transistor reliability and to the methods used to assure that they will function for the required time. Our discussion will assume bulk MOSFETs with SiO_2 gate dielectrics and heavily doped, polycrystalline Si gate electrodes — because most of the theory has been worked out for such devices — but the results remain relevant for modern, fully depleted, metal gate MOSFETs with high dielectric constant gate dielectrics.

The discussion of reliability often begins with the *bathtub curve* shown in Fig. 12.1, which is a plot of failure rate vs. time. It applies quite generally, not only to transistors, and is a superposition of two effects — the first is "infant mortality" failures caused by defects. So-called "killer defects" cause the device to fail immediately and are screened out during the initial testing. Other defects ("latent defects") introduced during the manufacturing process are not fatal but can become fatal. Examples include voids and pinholes in materials, resist residue, and particles. For critical applications, *burn-in testing* stresses devices under elevated voltages, temperatures, humidity, etc. and screens out the early failures. In modern silicon processes, defect densities are quite low so burn-in testing is not standard.

The second component in the bathtub curve is wear out — the accumulation of defects due to intrinsic processes that eventually cause failure of the transistors, or of the metal lines that connect the transistors, or of the package that they are enclosed in. In this lecture, our focus is on the intrinsic wear-out processes of Si MOSFETs. Three important wear-out processes, which all have to do with the gate oxide and the oxide-Si interface, will be discussed. The three processes are NBTI (Negative Bias Temperature Instability), HCD (Hot Carrier Degradation), and TDDB (Time Dependents Dielectric Breakdown).

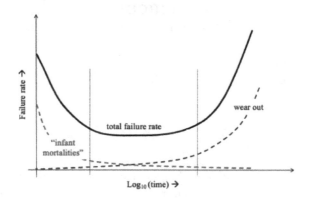

Fig. 12.1   The bathtub curve showing failure rate vs. time and the two components: 1) Early ("infant mortality") failures, which decrease with time, and 2) Wear-out failures, which increase with time.

Figure 12.2 is an atomic level sketch of an SiO_2 gate dielectric and the SiO_2:Si interface. The amorphous SiO_2 layer consists of Si-O bonds. At the surface of the Si, there are Si-O bonds and also a few Si *dangling bonds*. Below the surface are Si-Si bonds, but these are so strong that they are hard to break. Our focus, therefore, is on broken bonds at the SiO_2:Si interface and in the bulk of the oxide.

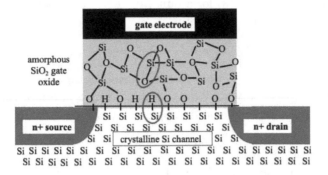

Fig. 12.2 A not-to-scale schematic illustration of an amorphous SiO$_2$ gate dielectric and the SiO$_2$:Si interface. Broken bonds can cause wear-out failure. Broken Si-H bonds at the interface and broken Si-O bonds within the oxide are especially important.

Dangling bonds

Early in the development of MOSFETs, it was learned that dangling bonds at the SiO$_2$:Si interface were electrically charged and could capture electrons from the channel and prevent MOSFETs from operating as intended. Oxygen-Si bonds at the SiO$_2$:Si interface tie up a large fraction of the dangling bonds, but a significant number remain. It was discovered, however, that hydrogen could *passivate* dangling bonds and render them electrically neutral. These are the Si-H bonds shown in Fig. 12.2. It was a combination of the high quality gate oxide that formed bonds with most of the Si atoms at the surface and the ability of hydrogen to passivate the remaining dangling bonds that made MOS technology viable, but over time the oxide and the interface can be damaged. When Si-H bonds at the interface are broken, Negative Bias Temperature Instability (NBTI) occurs. Energetic (so-called "hot") carriers can also break bonds at the interface and produce Hot Carrier Degradation (HCD). Broken Si-O bonds within the oxide can lead to Time Dependent Dielectric Breakdown (TDDB).

Defects and the threshold voltage

Both NBTI and HCD involve breaking Si-H bonds that cause the threshold voltage, V_T, to shift. To understand why, consider Fig. 12.2 and note that there are a few Si-H bonds, which, if broken, become positively charged. The result is a sheet of positive charge, qN_{IT} C/m^2, at the oxide-Si interface. Here, N_{IT} is the density of charged interface traps per m^2. How does this sheet of positive charge affect the threshold voltage?

Analogous to Eq. (8.13) for a bulk MOSFET, the relation between the gate voltage and charge in the Si is

$$V_G = -\frac{Q_S(\psi_s)}{C_{ox}} - \frac{qN_{IT}}{C_{ox}} + \psi_s \,,$$

where Q_S is the charge density in the semiconductor and we now include the interface charge, qN_{IT}. We see that a positive interface charge shifts the gate voltage, and therefore V_T, in the negative direction by

$$\Delta V_T = -\frac{qN_{IT}}{C_{ox}} \,, \tag{12.1}$$

so to understand how V_T shifts with time, we must understand how N_{IT} increases with time.

Reliability testing

Testing devices to ensure that they will survive for the required time presents challenges because the testing must be accomplished in a short time, but very long lifetimes must be guaranteed to the customer. Figure 12.3 illustrates the concepts. The vertical axis plots the degradation of key transistor parameters (e.g. the on-current or threshold voltage), and the horizontal axis is time on a log scale. We will see later that $N_{IT} \propto t^n$ (a power law in time), which becomes easy to read as a straight line on a log-log plot.

Because degradation occurs very slowly, *accelerated testing* is essential. The device is stressed under higher than normal voltages and/or temperatures (the solid lines labeled {1}), and from this accelerated testing data, one can deduce how the transistor will degrade under normal operating conditions (the dashed line {2}). Note that adjusting the accelerated test data {1} to the operating conditions, {2}, requires a detailed physical model. Data is only collected for about 10^5 seconds (about one day), but if we know the long time dependence of the degradation processes, we can extrapolate the degradation to much longer times, {3}, and determine whether the transistor has the required 10-year ($\approx 3 \times 10^7$ s) lifetime.

An equivalent way to project lifetime is to recognize that the equations that govern degradation have solutions that are a universal function of scaled time, $(S \times t)$,

$$\frac{dN_{\text{defect}}}{dt} = f(St) \,, \tag{12.2}$$

where N_{defect} is the concentration of defects responsible for the wear out, and $S(V_G, V_D, T, ..., \text{etc.}) > 1$ for accelerated testing. What this means is that accelerated testing simply acquires earlier data that occurs at longer

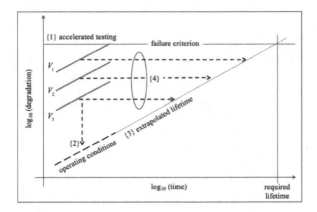

Fig. 12.3   Illustration of accelerated testing and extrapolated lifetime. The rate of degradation is increased by applying higher-than-normal voltages or temperatures (solid lines {1}). Using a detailed, physical model, the degradation rate under normal operating conditions can then be deduced (dashed line, {2}). The degradation vs. time is extrapolated to the failure criteria in order to deduce the extrapolated lifetime (line {3}). Alternatively, recognizing that degradation is a universal function of scaled time, we can simply translate the accelerated test data in time, {4} to produce line {3}.

times under normal operating conditions. So to predict degradation vs. time, we can simply translate the accelerated testing data horizontally in time (i.e. replace t by St) as shown by the dashed lines {4} in Fig. 12.3. The remarkable thing about this process is that it does not require a detailed physical model — only the knowledge that the degradation is a universal function of scaled time. It is important to understand that single parameter scaling works ONLY if there is a single physical phenomenon (e.g. NBTI) driving the degradation. Universality is broken when a combination of degradation processes occurs in parallel — in fact, this is a good way to check if stress has been over-accelerated and new phenomenon have been introduced. We will discuss both approaches.

12.2   Negative Bias Temperature Instability (NBTI)

The most important reliability issue for MOSFETs, NBTI, is a negative shift in the threshold voltage, V_T, of a p-channel MOSFET when the gate is biased in strong inversion (i.e. a negative gate to source voltage) [1-4]. This occurs for the PMOS transistor in the CMOS inverter of Fig. 2.1 when $V_{in} = 0$. An analogous effect but smaller in magnitude occurs for n-channel MOSFETs biased in strong accumulation (also a negative gate voltage). In both cases, there is a high concentration of holes at the interface, which gives a clue into the underlying physical mechanisms. The change in V_T produces a change in the linear region drain current (recall Eq. (4.5)), and the saturation region current (Eq. (4.7)). NBTI can also degrade the mobility, but our focus here will be on the threshold shift.

The NBTI V_T shift increases with time according to t^n and also depends on the electric field in the oxide, $\mathcal{E}_{ox}$, the temperature, T, and the frequency, f, and duty cycle, d, of the voltage applied to the gate. The V_T shift can be written phenomenologically as

$$\boxed{|\Delta V_T| = A(\mathcal{E}_{ox}) \times e^{-E_A/nk_BT} \times g(f,d) \times t^n,} \qquad (12.3)$$

where A is a function that describes the field dependence, E_A is the *activation energy* for the thermal processes, and g is a function that describes the frequency and duty cycle dependence.

Typical results for $\Delta V_T(t)$ are shown in Fig. 12.4. The PMOS transistor is stressed by applying a high voltage to the source and drain and grounding the gate. After stressing the device for a specified time (horizontal axis), V_T is quickly measured. As shown in Fig. 12.4, $|\Delta V_T|$ increases quickly with stress time, but after a second or so, it slows down and is described by Eq. (12.3) with $n \approx 1/6$. Remarkably, this power law behavior persists over 8–12 orders of magnitude in time. Our first goal is to explain why $|\Delta V_T(t)| \propto t^n$ and why $n \approx 1/6$.

Power law time dependence of NBTI

As indicated in Fig. 12.2, there are very few unpassivated Si bonds at the SiO_2:Si interface. A (100) Si surface has 6.8×10^{14} atoms/cm^2, but a good SiO_2:Si interface has on the order of 10^{10} charges/cm^2, so roughly one in every 100,000 Si atoms on the surface has a dangling bond. The low density of interface charges makes SiO_2:Si MOS technology possible, but when the device is stressed, some of the Si:H bonds break, which leads to a slow increase of N_{IT} with time. This is what causes NBTI.

The process of breaking a Si:H bond proceeds in three steps:

$$(Si - H) \rightarrow (Si - H)^+ \rightarrow Si^+ + H. \qquad (12.4)$$

In the first step, Si-H captures a hole (the stress bias ensures that there is a high concentration of holes at the interface), which leaves Si-H positively charged and reduces the energy needed to break the bond. Thermal energy then breaks the bond leaving a positively charged Si atom at the interface. In the third step, hydrogen diffuses away leaving the positively charged atom at the interface, which increases N_{IT}.

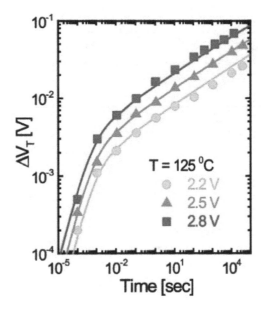

Fig. 12.4 Magnitude of the change in threshold voltage vs. stress time for a p-channel MOSFET stressed in strong inversion. (Data taken from H. Reisinger, O. Blank, W. Heinrigs, A. Muhlhoff, W. Gustin, C. Schlunder, 2006 IEEE International Reliability Physics Symposium Proceedings, 448–453, 2006.)

These processes are described with the *reaction-diffusion model*. The reaction is the breaking of Si-H bonds, and diffusion refers to the diffusion of hydrogen away from the interface. The reaction is described by

$$\frac{dN_{IT}}{dt} = k_F \left(N_0 - N_{IT} \right) - k_R N_H (x = 0) N_{IT} \,, \tag{12.5}$$

where N_0 is the starting areal density per m^2 of passivated dangling bonds, so $(N_0 - N_{IT})$ is the areal density of bonds not yet broken. The generation of interface states is proportional to the density of unbroken bonds, and the constant of proportionality is the forward disassociation rate, k_F. Hydrogen at the interface, $N_H(x = 0)$ m^{-3}, can react with broken bonds, N_{IT}, to repassivate the dangling bonds. This process is characterized by the reverse reaction rate, k_R.

The diffusion equation is

$$\frac{\partial N_{H_2}}{\partial t} = D_{H_2} \frac{\partial^2 N_{H_2}}{\partial x^2} \,, \tag{12.6}$$

where N_{H_2} refers to the concentration of molecular hydrogen, and D_{H_2} is the diffusion coefficient. Atomic H is released as bonds are broken, but H$_2$

quickly forms, and it is H_2 that subsequently diffuses. Together, these two equations are known as the reaction-diffusion (RD) model.

In general, numerical methods are needed to solve the reaction-diffusion equations, but simple, analytical solutions give insights. Traps are generated slowly. After a few seconds, each of the two factors on the RHS of Eq. (12.5) is large, but the difference is small, so, $dN_{IT}/dt \approx 0$. We'll also assume that $N_{IT} \ll N_0$; if N_{IT} is not small, the transistor will already have failed. With these two assumptions, Eq. (12.5) becomes

$$\left(\frac{k_F N_0}{k_R} \right) \approx N_H(x=0)N_{IT} \,. \tag{12.7}$$

To solve this equation for the trap density, we need the hydrogen concentration at the interface. For that, we turn to the diffusion equation.

Solutions to the diffusion equation show that diffusing particles move a distance of about $\sqrt{Dt}$ in a time, t. We can approximate the profile of $N_{H_2}(x)$ vs. x after a time, t, as shown in Fig. 12.5. Note that N_{H_2} is the concentration per m^3 of H in the SiO_2. The area under the $N_{H_2}(x)$ vs. x profile is the total amount of H_2 per m^2 in the film.

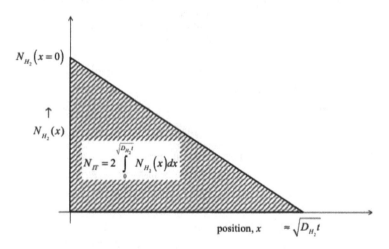

Fig. 12.5   Approximate profile of molecular hydrogen, H_2, as it diffuses in the SiO_2 film.

Recognizing that every H atom released in breaking a bond creates an interface trap, we see that each H_2 is responsible for two traps, so

$$N_{IT}(t) = 2\left(\frac{1}{2}N_{H_2}(x = 0)\sqrt{D_{H_2}t}\right). \tag{12.8}$$

In Eq. (12.7), we need the concentration of atomic hydrogen at the interface. How do we deduce $N_H(x = 0)$ from $N_{H_2}(x = 0)$ in Eq. (12.8)?

Two H atoms can combine to form an H_2 molecule, and an H_2 molecule can disassociate into to H atoms. In equilibrium, the reaction,

$$H + H \rightleftarrows H_2, \tag{12.9}$$

is described by the *law of mass action*

$$N_H \times N_H = KN_{H_2}, \tag{12.10}$$

where K is a constant. Using Eq. (12.10) for $N_{H_2}(x = 0)$ in (12.8), solving for $N_H(x = 0)$, and inserting the result in Eq. (12.7), we find

$$N_{IT}(t) \propto \left(\frac{k_F N_0}{2k_R}\right)^{2/3}(D_{H_2}t)^{1/6} = C \times t^{1/6}, \tag{12.11}$$

where $C = A(\mathcal{E}_{ox})e^{-E_A/nk_BT}g(f, d)$ in Eq. (12.3), and we have derived the proper time exponent, $n = 1/6$. The reaction-diffusion model explains why $N_{IT}(t)$ (and therefore, $|\Delta V_T|$) has a power law dependence on time. We leave it as an exercise to show that the diffusing species must be H_2 because if it were H, an incorrect exponent of $1/4$ would be obtained.

Frequency, duty factor, temperature, and e-field dependence

As indicated by Eq. (12.3), in addition to time, ΔV_T depends on the electric field in the oxide, $\mathcal{E}_{ox}$, the temperature, T, and the frequency, f, and duty cycle, d, of the voltage applied to the gate. Let's very briefly discuss the physics of these effects beginning with frequency and duty factor.

If the gate is driven by a clock, then the V_T shift with time is less than for a DC stress because when the stress voltage is removed, the Si-H bonds can re-form and N_{IT} will "relax" (decrease). Relaxation also affects measurements like those of Fig. 12.4; if the device is stressed for a given time after which the V_T shift is quickly measured, relaxation will occur and affect the measurement. Special techniques are needed to measure $\Delta V_T(t)$ accurately (Lectures 17–20 in [5]).

Careful analysis shows that if the device is stressed for a long time, t, then the total V_T shift depends only on the total time that the device was stressed, $d \times t$; it is independent of frequency (at least up to about 2 GHz).

As a result, duty factor corrections can be made to deduce the TTF for operating circuits from the measured DC threshold voltage shifts.

The breaking and forming of Si-H bonds are thermally activated, but the temperature dependence of k_F and k_R are similar, so according to Eq. (12.11), these processes do not contribute the temperature dependence of ΔV_T. In contrast to the diffusion of electrons and holes in crystalline semiconductors, the diffusion of H_2 in amorphous SiO_2 has a strong temperature dependence described by

$$D_{H_2} \propto e^{-E_{H_2}/k_B T} . \tag{12.12}$$

This strong temperature dependence arises from defects like oxygen vacancies that produce deep potential wells that trap H_2. Higher temperatures lead to higher thermal energies that reduce the time that H_2 spends in these traps and, therefore, increases D_{H_2}.

By using Eq. (12.12) in Eq. (12.11), we find

$$N_{IT}(t) \propto e^{-E_{H_2}/6k_B T} , \tag{12.13}$$

so, the activation energy for $|\Delta V_T|$ in Eq. (12.3) is the activation energy of H_2 diffusion, $E_A = E_{H_2} \approx 0.58$ eV.

As shown in Fig. 12.4, $|\Delta V_T|$ increases with stress voltage. Careful analysis actually shows that $|\Delta V_T|$ depends on the electric field in the oxide, $\mathcal{E}_{ox}$, which is controlled by the gate voltage. The field dependence of N_{IT} is due to the field dependence of k_F in Eq. (12.11); k_R and D_{H_2} are independent of the field in the oxide. We can write the field dependence as

$$k_F \propto Q_p(\mathcal{E}_{ox}) \times \mathcal{T}_p(\mathcal{E}_{ox}) \times e^{-E_B(\mathcal{E}_{ox})/k_B T} , \tag{12.14}$$

where Q_p is the inversion layer charge in C/cm^2. Holes are needed at the interface because when holes are captured by Si-H, the bond breaking energy, E_B, is lowered. Assuming that the interface charge is low, then Q_p is directly related to $\mathcal{E}_{ox}$ by Gauss's Law: $Q_p = \kappa_{ox}\epsilon_0\mathcal{E}_{ox}$.

Holes must tunnel through an energy barrier at the interface to be captured by an Si-H bond. The $Si : SiO_2$ energy barrier is about 4.7 eV, but the thickness of the triangular barrier is determined by the electric field in the oxide. This is the Fowler-Nordheim tunneling that we discussed briefly in Sec. 10.6, so the hole transmission, $\mathcal{T}_p$, is a function of $\mathcal{E}_{ox}$.

Finally, $\mathcal{E}_{ox}$ lowers the bond breaking energy, E_B, because the Si-H bond has a dipole moment; the H is slightly negative and the Si slightly positive, so the electric field reduces the dipole moment and lowers E_B.

Reliability testing

Mathematical models for $k_F(\mathcal{E}_{ox})$ can be used to determine $A(\mathcal{E}_{ox})$ in Eq. (12.3), which then provides a complete model for the accelerated testing of NBTI. Damage occurs too slowly to gather data under normal operating conditions, so we increase the gate voltage or temperature, acquire accelerated data for about one day, then use Eq. (12.3) to translate this data to the operating conditions, and finally extrapolate in time as $t^{1/6}$. Alternatively, as discussed with respect to Fig. 12.3, we could just scale time in the accelerated data and produce a $\Delta V_T(t)$ characteristic for long time operation under normal operating conditions.

Complicating factors

We have discussed the main features of NBTI, but accurate prediction of NBTI lifetimes requires a deep understanding of the relevant physics and careful attention to details. One complicating factor is that in addition to NBTI, other processes may shift V_T. An example is hole trapping in the oxide; mobile holes in the PMOS channel can be injected into the oxide and trapped. These trapped positive charges shift V_T. The measured data must be corrected for hole trapping so that accurate projections of NBTI lifetimes can be made. Finally note that anomalous exponents, $n \neq 1/6$ can be observed. This occurs, for example, in "nitrided" SiO_2, which has a high nitrogen concentration. See Lectures 8–12 in [5] for a more detailed and more comprehensive treatment of NBTI.

12.3   Hot Carrier Degradation (HCD)

When stressing MOSFETs to characterize NBTI, $V_{DS} = 0$. Charge carriers in the channel are in thermal equilibrium with an average energy of $3k_BT/2$, where T, the temperature of the semiconductor lattice, is also the temperature of the charge carriers because the two systems are in thermal equilibrium. When $V_{DS} > 0$, however, the carriers gain energy as they move from source to drain. We write their energy as $3k_BT_e/2$, where T_e is the *electron temperature*. Electrons gain energy more rapidly than they shed it to the lattice, so $T_e > T$, and we refer to these energetic carriers as *hot carriers*. The carriers are most energetic near the drain where they can cause damage that results in *hot carrier degradation* [6–9].

Early MOSFETs used $V_{DD} = 5$ V to be compatible with the then dominant technology — bipolar technology. As channel lengths shrunk, electric fields increased, and HCD became a serious reliability issue. When the lateral electric field is high, *impact ionization* can create electron-hole pairs, or carriers may scatter and be injected into the oxide where they may contribute to a gate current, be trapped in defects in the oxide, or create new defects in the oxide. Special structures (so-called *lightly doped drains*, LDDs), analogous to the extended drains discussed for power MOSFETs in Sec. 9.2, were developed to reduce the electric field and manage HCD.

For modern logic transistors with $V_{DD} < 1$ V, HCD is still a serious reliability issue. The HCD of most concern is caused by the breaking of Si:H and Si:O bonds at the SiO_2 : Si interface near the drain end of the channel, which requires more than 1 eV of energy. Where does this energy come from when $V_D \approx 1$ V? One possibility is that the bonds are broken in a series of small thermal excitations. Another possibility is that carrier-carrier scattering produces high energy tails in the carrier distribution, even though the average energy is much lower.

Broken bonds create interface charges that affect the performance of transistors. The hot carrier V_T shift can be written phenomenologically as

$$\boxed{|\Delta V_T| = A(V_{DS}) \times e^{-E_A/k_B T} \times g(f) \times f(t)\,.}$$ (12.15)

In comparison to Eq. (12.3) for NBTI, HCD does not depend on the electric field in the oxide, $\mathcal{E}_{ox}$, but it does increase strongly with V_{DS}. There is an activation energy that describes the temperature dependence, and it depends on frequency, but not on the duty factor. Finally, as shown in Fig. 12.6, HCD degradation follows a power law, but the characteristic exponent varies between about $n \approx 1/4$ to $n \approx 1/2$. HCD degradation also shows a long term, "soft saturation" that complicates the determination of the time to failure, TTF. This soft saturation is caused by the fact that there is dispersion in the bond-breaking energy, which depends on the orientation of the bond and its local environment. Another key difference between HCD and NBTI is that when the stress voltage is removed, there is relatively little relaxation of the damage. We seek to understand HCD so that we can test transistors and project their lifetime.

Time dependence of HCD

For NBTI, we were able to explain the $t^{1/6}$ behavior using simple arguments. What is the expected t^n dependence when Si:H bonds are broken by hot carriers? We begin by assuming that bond-breaking and bond forming processes in Eq. (12.5) are each large, so that the difference is small such that

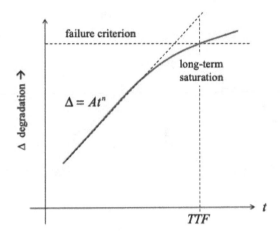

Fig. 12.6 Sketch of typical degradation vs. time for hot carrier degradation (HCD). The degradation may be a change of V_T, I_{Dlin}, I_{Dsat}, or g_m — all show similar characteristics.

$dN_{IT}/dt \approx 0$ resulting in Eq. (12.7) again. The difference between NBTI and HCD is in the diffusion of H_2 after the bonds are broken. As shown in Fig. 12.7, diffusion is one-dimensional for NBTI but two-dimensional for HCD.

The number of traps is equal to the total amount of hydrogen that has diffused away from the damaged region near the drain. If we assume that H_2 diffuses radially from a line source at the end of the channel, we find

$$N_{IT} A_d = 2 \times \frac{\pi \left(\sqrt{D_{H_2} t} \right)^2}{4} W N_{H_2}(x = 0, y = L).$$ (12.16)

where N_{IT} is the trap density per m^2, W the width of the transistor and A_d is the area of the damaged region near the drain. The RHS is the total amount of H that was released in bond-breaking with $N_{H_2}(x = 0, y = L)$ being the concentration per m^{-3} in the damaged region at $x = 0$ and near $y = L$. As discussed for NBTI, the law of mass action gives $N_{H_2} \propto N_H^2$, which can be inserted in Eq. (12.16), solved for $N_H(x = 0)$ and the result inserted into Eq. (12.7) to find

$$N_{IT}(t) \propto \left(\frac{k_F N_0}{2k_R} \right)^{2/3} (D_{H_2} t)^{1/3} = C \times t^{1/3},$$ (12.17)

which should be compared to Eq. (12.11) for NBTI. The time-dependent exponent has increased from $n = 1/6$ to $n = 1/3$. The assumption of a

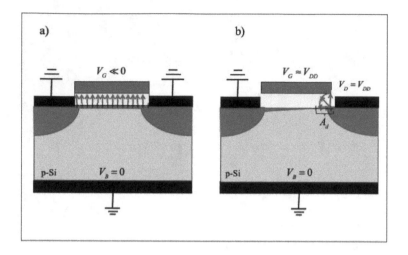

Fig. 12.7   Illustration of the difference in the diffusion of H_2 in NBTI testing, where it is one-dimensional and in HCD, where is it two-dimensional.

line source generation of defects at the drain end of the channel makes this a highly simplified analysis, but it helps to explain why $n \approx 1/4 - 1/2$ are commonly observed for HCD instead of the $n \approx 1/6$ for NBTI.

The analysis of degradation vs. time for HCD is complicated by the decrease of n due to the long-term saturation behavior. This behavior is attributed to the fact that there is a distribution of bond energies, which becomes important for hot carriers because of the wide range of energies involved. Moreover, detailed numerical solutions of the RD model show that if the length of the damaged region near the drain increases with drain voltage, then the exponent, n decreases with V_{DS}. These factors mean that we cannot analyze HCD data with a simple power law dependence on time, but our simple solution of the RD model does provide qualitative insights. For example, the geometry of diffusion explains why there is much less relaxation in HCD than in NBTI when stress voltage is removed. As shown in Fig. 12.7, when the NBTI stress is removed and bonds are no longer being broken, the hydrogen can diffuse back to the interface and anneal the damage, but for HCD, much less of the H_2 will find its way back to the small damaged region.

Frequency and temperature dependence

For long channel MOSFETs, maximum HCD occurs at $V_{GS} \approx V_{DS}/2$ because as V_{GS} increases, the density of carriers in the channel increases, so there are more energetic carriers to cause damage, but as V_{GS} approaches V_{DS}, the transistor leaves the saturation region, and there is no longer a pinch-off region with a high electric field near the drain. For short channel MOSFETs, however, maximum HCD occurs at $V_{GS} \approx V_{DS}$, because the electric field remains high.

The switching frequency dependence of HCD is distinctly different in long and short channel MOSFETs. For long channel MOSFETs, HCD increases with frequency because for CMOS logic, the $V_{GS} \approx V_{DS}/2$ condition occurs during the switching transition, and the number of these is proportional to frequency. For short channel MOSFETs, self heating dominates. The switching frequencies are much higher than the thermal response of the device, so the temperature rise during a period is limited. The higher the frequency, the lower the peak temperature, and the peak temperature is what's important for bond-breaking. The result is less HCD as the frequency increases.

The lattice temperature dependence of HCD is a balance of two factors. An increase in T increases the density of phonons (lattice vibrations), which interact with electrons and lower their energy (temperature, T_e) by providing more channels for electrons to dissipate their energy to the lattice. This factor dominates for long channel transistors and causes the time to failure, TTT, to increase as T increases. The second factor, the increase in the bond dissociation probability at higher temperatures,

$$k_F(T) \propto e^{-E_B/k_B T}, \qquad (12.18)$$

prevails for short-channel MOSFETs because to control short-channel effects, modern transistor structures surround the channel with low thermal conductivity gate dielectrics (see the FinFET and nanosheet transistors in Fig. 3.11), which makes it hard for the heat generated in the channel to escape. The result of this *self-heating effect* is to greatly increase T, which causes the TTF to decrease as T increases. The MOSFET structures developed to control short-channel effects have greatly exacerbated HCD.

Reliability testing

Figure 12.8 shows HCD for a $V_{DD} = 1$ V technology. For reasonable testing times, very little HCD occurs, but if we increase V_{DD}, HCD greatly increases, and data can be collected. How can we use the data at higher

V_{DD}'s to project the HCD lifetime at the operating voltage? For NBTI, we translated the higher voltage data to the operating voltage and then extrapolated in time according to $t^{1/6}$. We cannot do this for HCD because the time dependence is not a simple power law.

NBTI is described by the reaction-diffusion model, Eqs. (12.5) and (12.6). HCD is described by a generalization to a distribution of bond energies. The remarkable thing is that the solutions to equations of this kind are a universal function of scaled time as given by Eq. (12.2). As shown in Fig. 12.8 by translating the higher V_D results in time (i.e. replacing t by St on the log plot) the results fall on a universal degradation curve. This procedure allows us to generate a degradation vs. time characteristic under operating conditions that extends to long times and large amounts of degradation. One can then just read the TTF off the universal curve.

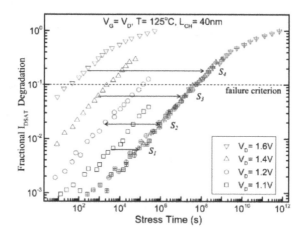

Fig. 12.8   Measured HCD vs. time for five different drain voltages. Also shown is how by scaling time for each of the accelerated tests, a universal degradation function is obtained. (©IEEE 2018. Reprinted, with permission, from [6].)

Universal scaling is broadly applicable to a variety of experimental conditions such as different lattice temperatures, frequencies, channel lengths, number of FinFET fins, etc. [6]. The various stress conditions trace different parts of a single, universal degradation characteristic. Using the scaling function, $S(V_D, V_G, T, ...)$, one can project HCD to arbitrary combinations of stress conditions. The scaling factors can be obtained from limited time measurements, which provides an easy and accurate method for lifetime extrapolation. Note that recognizing HCD results in a universal degradation characteristic vs. time allows us to project the lifetime without understanding and mathematically modeling the underlying physics. NBTI lifetimes can also be determined in the same model-free way.

12.4   Time Dependent Dielectric Breakdown (TDDB)

Dielectrics fail by breaking down and shorting two conductors together. Like NBTI and HCD, dielectric breakdown is caused by defects, but dielectric breakdown is caused by defects in the bulk of the oxide, not at the oxide-Si interface. Dielectric breakdown is fundamentally different for the relatively thick dielectrics that separate the metal layers of an integrated circuit and for the very thin gate dielectrics used in MOSFETs. Our focus here will be on breakdown in thin, gate dielectrics.

Figure 12.9 illustrates how defect generation leads to breakdown in a thin dielectric and sketches the typical gate current vs. time. As shown on the left, defects are generated at the interface and in the bulk of the gate dielectric. Defects at the interface are responsible for NBTI and HCD, and defects in the bulk facilitate the transport of electrons across the dielectric and therefore, slowly increase the gate current. As shown on the right, this is *Stress Induced Leakage Current* (SILC). As also shown in Fig. 12.9a, the random defect generation process can eventually produce a conducting path across the dielectric. When this happens, there is an abrupt increase in current, but if the dissipated power is below a critical level, no permanent damage is done and the transistor continues to operate. This is known as *soft breakdown*, but eventually, *hard breakdown* occurs, and the transistor no longer functions.

Physics of TDDB
The *Anode Hole Injection* (AHI) model describes TDDB in both thick and very thin gate dielectrics [10, 11]. Figure 12.10 summarizes how defects in the oxide are generated according to the AHI model. For an n-channel MOSFET biased in inversion, the process begins with electrons tunneling from the cathode (the channel) to the anode (the gate electrode, shown here as heavily doped, n^+ polycrystalline Si). Note that for a thin oxide, this is *direct tunneling*, not Fowler-Nordheim tunneling because the tunneling thickness is determined by the film thickness, not by the electric field. Electrons arrive in the anode with a kinetic energy that is determined by

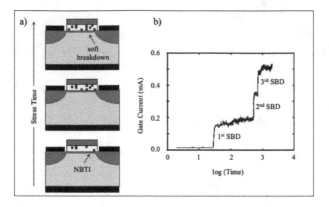

Fig. 12.9   a) Illustration of how defects in the gate dielectric lead to breakdown. b) Typical gate current vs. time characteristics. (Figure 12.9b provided by M.A. Alam, Purdue Univ., 2022.)

the gate voltage.   If the electron energy is high enough, electrons may collide with electrons deep in the valence band of the gate electrode, break a covalent bond, lose their kinetic energy and drop to the bottom of the conduction band. In this *impact ionization* process, an electron-hole pair is generated by moving an electron deep in the valence band to an empty state near the bottom of the conduction band, leaving behind energetic holes in the valence band. The energetic holes ("hot holes") can tunnel back through the oxide, and in the process break bonds and create defects in the oxide. This process continues until the number of defects reaches a critical value, N_{BD}, and breakdown occurs. The oxide defect generation process can be described phenomenologically as

$$N_{\text{defect}} = (J_p k)t = (J_n \alpha_{ii} \mathcal{T}_p k)\, t \,, \qquad (12.19)$$

where J_p is the hole tunneling current through the oxide, and k is the efficiency at which they generate traps. The hole current is the product of J_n, the electron tunnel current, α_{ii}, the probability that electrons impact ionize in the gate electrode (anode), and the transmission, $\mathcal{T}_p$, which describes the tunneling of hot holes through the oxide.

Breakdown occurs when a critical density of traps is generated, so

$$T_{BD} = \left(\frac{N_{BD}}{k}\right) \frac{1}{J_p} \,. \qquad (12.20)$$

The impact ionization efficiency, α_{ii}, is strongly dependent on the energy of the electrons that tunnel into the anode, and this is determined by the gate

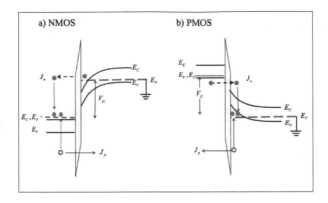

Fig. 12.10   a) Illustration of how the AHI model explains the creation of defects in the oxide of an n-channel MOSFET in strong inversion. b) Oxide defect generation in the AHI model for a p-channel MOSFET biased in strong inversion. To keep things simple, the small band bending in the polysilicon gates is ignored.

voltage. The reduction of power supply voltage to manage power dissipation in ICs greatly lowered the probability of impact ionization, which lowered J_p and dramatically improved the reliability of thin gate oxides.

Figure 12.10b shows how defects are created in p-channel transistors biased in strong inversion. The process begins with electrons tunneling from the cathode (the valence band of the gate electrode) to the anode (the inversion layer of holes in the n-type semiconductor). Electrons arrive in the anode with a kinetic energy that is determined by the gate voltage. In contrast to the n-channel case, electrons may drop all the way down to empty states in the valence band of the semiconductor in strong inversion. The result is that the electron energy available to impact ionize is very high, so very hot holes are generated. These energetic holes find it easier to tunnel through the oxide, so J_p, which is what creates defects in the oxide, is higher for p-channel MOSFETs than for n-channel devices. *The result is that TDDB is a much more serious issue for PMOS transistors than for NMOS transistors.*

Statistical theory of TDDB

Let's turn now to the statistics of breakdown. When defects are created within the oxide, the leakage current increases because electrons can hop (*percolate*) from defect to defect. As shown in Fig. 12.9, when a complete path (a *percolation path*) is produced between the inversion layer and the gate electrode, current flows and soft breakdown occurs. Let's assume that the gate oxide is divided into cubes of side a, which is the size of a defect. For SiO$_2$, this is a broken Si-O bond and $a \approx 3$ Å. Assuming that there are M cubes in the thickness of the oxide and that the area of the gate contains N cubes, we find

$$M = t_{ox}/a$$
$$N = A_G/a^2 \,. \tag{12.21}$$

Let $q(t)$ be the probability that a defect has been created before time, t, and let's compute the probability that no percolation paths have occurred up to the time, t, assuming that the defects are uncorrelated. At any location across the area of the gate, the probability of a percolation path is

$$p(t) = q(t)^M \,. \tag{12.22}$$

The probability that a percolation path has not been created is $(1 - p(t))$, and the probability that no path has been created anywhere across the area of the gate oxide is

$$(1 - p(t))^N \,. \tag{12.23}$$

Now let $F(t)$ be the *cumulative distribution function*, the fraction of devices that have failed by time, t. The probability the device has not failed (i.e. that there are no percolation paths) is $1 - F(t)$, so from Eq. (12.23), we see that

$$1 - F(t) = (1 - p(t))^N = (1 - Np/N)^N = e^{-Np} \,, \tag{12.24}$$

where we assumed that N is large and made use of

$$\lim_{n \to \infty} (1 - x/n)^n = e^{-x} \,.$$

The cumulative distribution function is easy to measure; we stress a small number of devices by grounding the source and drain and applying a gate voltage that is above the threshold voltage. We then keep track of the fraction that fail vs. time, that is $F(t)$. In practice, we don't plot $F(t)$ or $1 - F(t)$, but, rather, the *Weibull function*, $W(t)$ where

$$W(t) \equiv \ln\left[-\ln(1 - F(t))\right] = \ln(Np) = \ln(p) + \ln N \,. \tag{12.25}$$

Now let's look more closely and put some physics in.

We will assume that the probability that a defect has been generated follows a power law in time:

$$q(t) = (t/t_0)^\alpha \,, \tag{12.26}$$

where t_0 is a constant and α is the characteristic exponent (recall that $\alpha = 1/6$ for NBTI defect generation). The probability that a percolation path is created is

$$p = q^M = (t/t_0)^{M\alpha} = (t/t_0)^{\beta} \tag{12.27}$$

where

$$\beta \equiv M\alpha = t_{ox}\alpha/a \tag{12.28}$$

is proportional to the oxide thickness. Using Eq. (12.27) in (12.25), we find

$$\boxed{W(t) = \beta \ln(t) - \beta \ln(t_0) + \ln(A_G/a^2),} \tag{12.29}$$

where we have also used Eq. (12.21) to relate N to the gate area. Equation (12.29) is an important result; it says that if we measure $F(t)$, and then plot the corresponding $W(t)$ vs. $\ln(t)$, the result is a straight line with a slope of β. The *Weibull slope*, β, is a very important parameter.

Reliability testing

The typical reliability requirement for ICs is that the probability that a chip will fail within 10 years must be less than 0.01%. Failure statistics can be collected only on a small number of test transistors, and because the breakdown processes are slow the testing must be accelerated in order to collect statistically significant data. This is usually done by increasing the gate voltage. We then need a theory to deduce the Weibull plot at the operating voltage from the data obtained at higher voltages. The total gate area of the devices tested will be much less than the total gate area of the transistors on the IC, so we also need theory to deduce the Weibull plot for the appropriate gate area. Knowing the Weibull plot for the chip under standard operating conditions, the failure probability after 10 years can be read off the plot.

Figure 12.11 shows how this all comes together. Recall that according to the AHI model, the probability that hot holes will be generated by electrons that initiate impact ionization depends exponentially on the electron energy, which is a function of the gate voltage. The result is that the rate of defect generation is exponentially related to the gate voltage. To collect failure data in reasonable times, we apply a gate stress voltage that is greater than V_{DD}, the operating voltage. Figure 12.11 (lines {1}) shows Weibull plots for three such test voltages: V_1, V_2, V_3. As expected from Eq. (12.28), the Weibull slope is independent of voltage — so these

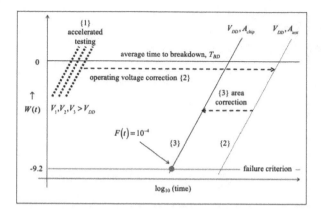

Fig. 12.11   Illustration of how accelerated TDDB testing works. First, Weibull plots at three test voltages $V_1, V_2, V_3 > V_{DD}$ are produced by testing a small number of samples (lines {1}). Next, we scale these results to the operating voltage, V_{DD}, to produce line{2}. Finally, we correct for the fact that the total gate area of the chip is much greater than that of the tested devices, to produce line {3} from which the reliability of the chip can be predicted from the Weibull slope.

measurements allow us to determine the Weibull slope — but we need a Weibull plot at the lower, operating voltage of V_{DD}.

It should take exponentially longer to produce a specific $F(t)$ (or $W(t)$) at a lower voltage, so we can write

$$t_{V_2} = t_{V_1} e^{\gamma_v \Delta V} , \tag{12.30}$$

where γ_v is the *voltage acceleration factor*, and ΔV is the difference in voltages. By testing three devices at high voltages, we can determine β and γ_v. To produce a Weibull plot at the operating voltage, we use Eq. (12.30) to shift the measured data in time according to

$$\log(t_{V_{DD}}) = \log(t_{V_{test}}) + (V_{test} - V_{DD})\gamma_v . \tag{12.31}$$

This *voltage scaling* produces line {2} in Fig. 12.11.

One more correction is needed to predict the reliability of a chip. Only a small number of test transistors (on the order of 100) is used, and the total gate area of these test transistors is much less than the total gate area of the chip. Because of this, the chip will fail much earlier, so we must make an area correction to obtain the correct Weibull plot. By equating $W(t)$ for two different gate areas, we find that the time for the IC to fail,

t_{IC}, is related to the time for curve $\{2\}$, $t_{[2]}$, according to

$$t_{IC} = t_{[2]} \left(\frac{A_G(\text{test})}{A_G(\text{IC})} \right)^{1/\beta}$$

$$\log(t_{IC}) = \log(t_1) + \log \left(\frac{A_G(\text{test})}{A_G(\text{IC})} \right)^{1/\beta},$$

(12.32)

which means that failures of the chip occur much earlier in time than for the test transistors. This area correction produces line $\{3\}$ in Fig. 12.11.

We can now determine the reliability of the IC. No more than 100 chips out of 1 million (i.e. 0.01 %) may fail within 10 years, so

$$F(t = 10\,\text{years}) = 10^{-4}$$
$$W(F = 10^{-4}) = -9.2\,.$$

Since line $\{3\}$ in Fig. 12.11 is the Weibull plot at the operating voltage and for the correct chip gate area, we simply read off the time at which $W = -9.2$ occurs and see if it at least 10 years.

As MOSFET device engineers scaled channel lengths down to increase the number of transistors on a chip and stay on the Moore's law trajectory, the gate oxide thickness was continually reduced. As discussed in Sec. 3.9, this was necessary to maintain strong gate control of the source to channel energy barrier, which was needed to manage short channel effects. Engineers became gravely concerned with the reliability of these thin gate oxides because, as Eq. (12.28) shows, the Weibull slope β is proportional to the oxide thickness. A shallower Weibull slope would make it harder and harder to achieve the required reliability. It turns out, however, that the theory we have outlined is overly pessimistic.

To understand why our theory is overly pessimistic, recall Fig. 12.9. A soft breakdown is not fatal if the power dissipation is limited. As gate oxide thicknesses were scaled down, so was the voltage, so transistors could survive multiple soft breakdown events with little change in I_D or g_m. The key is that breakdown in thin oxides is uncorrelated. If a small transistor on a chip has suffered a soft breakdown, its chances of suffering a second or third soft breakdown do not increase. For any given transistor that has suffered a soft breakdown, it is very unlikely that it will suffer another soft breakdown. If it takes multiple soft breakdowns to degrade the transistor performance enough to cause the circuit to fail, then we need to make another correction to line $\{3\}$ in Fig. 12.11 and correct for the probability that multiple soft breakdowns will occur. The statistics of multiple breakdowns are only a little more involved that the first breakdown statistics and are discussed in the Lecture 12 Exercise. The bottom line is that if transistors can survive n soft breakdowns, then the corresponding Weibull slope is $\beta_n = n\beta_1$, where β_1 is the Weibull slope for one soft breakdown as given by Eq. (12.28). The fact that transistors can survive multiple soft breakdowns dramatically increases their lifetimes and makes modern, high-performance CMOS possible.

12.5   Discussion

Data like that shown in Fig. 12.4 for NBTI and in Fig. 12.8 for HCD is collected from large area test structures and, as such, represents the average behavior of many small transistors. Data like this is used to predict the average time to failure due to NBTI and HCD. For TDDB, however, we argued that knowing the average time to breakdown, T_{BD}, was not sufficient; we needed a statistical theory to be sure that there are no more than 0.01% failures in the first 10 years of operation. Why are averages sufficient for NBTI and HCD but not for TDDB?

The basic reason for this difference is that there is some averaging that occurs as V_T shifts due to NBTI and HCD, but breakdown is an all or nothing event — a transistor has either failed or not. During NBTI and HCD degradation, the degradation is driven by the gate over drive, $(V_{GS} - V_T)$. If the V_T of a transistor is less than the average, the stress is greater, so the degradation is faster than average. If V_T is higher than average, the stress is less so the degradation is slower than average. The result is that the distribution of $V_T's$ moves toward the mean. Another type of averaging occurs in the circuits. A critical delay path will involve a series of logic gates. Higher $V_T's$ lower the drive current and reduce the speed, and lower $V_T's$ increase the drive current and increase the speed. The result is that the delay in the critical path depends on the average V_T in the critical path. In contrast, when one transistor fails due to TDDB, no averaging is possible; the circuit fails. Average time to breakdown is not relevant to TDDB reliability.

Electronic systems can fail for many different reasons; we have only discussed three, intrinsic wear-out processes associated with the gate oxide of a MOSFET. The metal wires that interconnect transistors are also subject to a wear-out process called *electromigration*, which is the diffusion of atoms and vacancies in a metal line driven by the momentum imparted by the electrons flowing through the wire. A divergence of the atom or vacancy flux can occur, for example, at grain boundaries and produce voids that open conductors or hillocks that short conductors. The mean time to failure is described empirically by *Black's equation*:

$$MTTF = CJ^{-n}e^{-E_A/k_B T}, \qquad (12.33)$$

where C is a material-dependent constant, J the current density, $1 < n < 2$ an exponent that describes the current dependence, and E_A the material-dependent activation energy. Electromigration is very material dependent; for example, copper is about five times more resistant to electromigration than aluminum. Electromigration is also very sensitive to the current density and to the temperature of the metal line. It can be an issue in the upper level metal lines, which are long and carry dc currents. The ac current in the first metal layer (M1), should reduce electromigration, but the metal lines in this level are heated by the self-heating of the transistors, which aggravates electromigration.

Failures can also occur in the chip packages due to thermo-mechanical stresses and corrosion and in the boards that the packages sit on. Electronics in space is subject to radiation, which can lead to transient or permanent damage of transistors. There are many modes of failure, but careful attention to NBTI, HCD, and TDDB as well as electromigration ensures that these intrinsic processes do not limit the reliability of the overall system.

12.6 Summary

This lecture has touched on only a few, important, intrinsic reliability processes for MOSFETs. Science-based reliability requires a deep understanding of complex physics that must be translated into simple models that can be used to analyze experimental data. We discussed three examples: Negative Bias Temperature Instability (NBTI), Hot Carrier Degradation (HCD), and Time Dependent Dielectric Breakdown (TDDB).

NBTI is an important issue for p-channel MOSFETs because they experience a negative gate-to-source bias in CMOS circuits. NBTI is described by a power law dependence on time that can be explained with the reaction-diffusion model and used to estimate the average time to failure.

HCD affects both n-channel and p-channel MOSFETs and continues to be a reliability concern even in modern, low-voltage logic transistors. The time universality of HCD permits us to estimate the average time to failure.

TDDB is now mainly an issue for p-channel MOSFETs because the Anode Hole Injection (AHI) process produces very hot holes, which cause the damage in the oxide that leads to breakdown. The fact that soft breakdown occurs in thin gate oxides and because transistors continue to function even with multiple soft breakdowns dramatically extends the lifetime. Because the failure of one transistor due to gate breakdown will cause the circuit to fail, we cannot guarantee reliability from the average time to breakdown. Instead, a statistical theory is used to determine the probability of failure (typically 0.01%) after a prescribed time (typically 10 years).

For a more comprehensive treatment of NBTI, HCD, and TDDB as well as an introduction to radiation effects, and the important topic of collecting and analyzing data, see the online lectures by M.A. Alam [5].

Lecture 12 Exercise: Percolation theory of multiple soft breakdowns

According to Eq. (12.24), the probability that a transistor will survive for a time t with no percolation paths is

$$P_0(t) = 1 - F(t) = \exp(-Np) = \exp(-N(t/t_0)^\beta).$$

Let's compute the probability, P_n, that a transistor will survive for a time t with $n-1$ or fewer percolation paths. The answer is

$$P_n = p^n \times (1-p)^{N-n} \times {}^N C_n,$$

where p^n is the probability of n percolation paths being formed, $(1-p)^{N-n}$ is the probability that there are no percolation paths in the remaining $(N-n)$ columns, and since we don't care where within the transistor the n percolation paths occurred, we multiply by the combination of N columns taken n percolation paths at a time. By expanding the terms in P_n, we find

$$P_n = p^n \times (1 - Np/N)^{N-n} \times \frac{N!}{(N-n)!n!}$$

$$= p^n \times (1 - Np/N)^{N-n} \times \frac{N(N-1)...(N-n+1)}{n!}$$

$$\approx p^n \times (1 - Np/N)^N \times \frac{N^n}{n!},$$

where in the last line we have made use of the fact that $N \gg n$. Making use of the fact that $(1 - Np/N)^N \approx \exp(-Np)$ as we did in Eq. (12.24), we find

$$P_n \approx \frac{(Np)^n}{n!} \times e^{-Np}.$$

Note that if we set $n = 0$, we recover the expression for P_0 given above.

The quantity $P_n(t)$ is the probability that exactly n soft breakdowns occur in the transistor by time, t. The probability that n or more soft breakdowns will occur by time, t, is

$$F_n = 1 - \sum_{k=0}^{n-1} P_k.$$

It can be shown from this result that

$$1 - F_n \propto \frac{(Np)^n}{n!}$$

(see the Appendix of Lecture 24 in [5]). Finally, we can find the corresponding Weibull distribution for n defects in a transistor from the above result and

$$W(t) \equiv \ln\left[-\ln(1 - F(t))\right] .$$

The result is

$$\boxed{W_n(t) = n\beta_1 \ln(t) + \text{constant} ,}$$

where β_1 is the Weibull slope for $n = 1$ soft breakdowns as given by Eq. (12.28). This result should be compared to the corresponding expression for $W_1(t)$ as given by Eq. (12.29). The important point is that the Weibull slope scales with the number of soft breakdowns.

The figure below summarizes the implications. For one soft breakdown ($n = 1$), the time to failure is quite short, but for $n = 2$ soft breakdowns, it is orders of magnitude larger, and for $n = 3$, it is even larger. The fact that thin oxides break down softly and that transistors continue to function with multiple soft breakdowns provides a remarkable enhancement of the lifetime without which p-channel MOSFETs would not be sufficiently reliable.

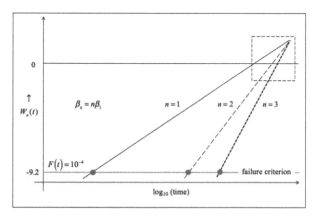

The n-soft breakdown Weibull plots for $n = 1$, 2, and 3 showing how the Weibull slope becomes steeper and the projected lifetime longer when multiple soft breakdowns are allowed. The box in the upper right shows where the data is collected. Accurate Weibull slopes much be deduced to permit extrapolation to very low failure probabilities.

12.7 References

NBTI has been a reliability concern since the 1970s. The understanding of NBTI physics and the modeling of it has evolved along with transistor technology. The following papers summarize this evolving understanding of NBTI.

[1] M.A. Alam and S. Mahapatra, "A comprehensive model of PMOS NBTI degradation," *Microelectronics Reliability*, **45**, pp. 71–81, 2005.

[2] M.A. Alam, K. Kufluoglu, D. Varghese, and S. Mahapatra, "A comprehensive model of PMOS NBTI degradation: Recent Progress," *Microelectronics Reliability*, **47**, pp. 853–862, 2007.

[3] A.E. Islam, H. Kufluoglu, D. Varghese, S. Mahapatra and M. Alam, "Recent issues in negative-bias instability: Initial degradation, field dependence of interface trap generation, hole trapping effects, and relaxation," *IEEE Trans. Electron Dev.*, **54**, pp. 2143–2154, 2007.

[4] S. Mahapatra, N. Goel, S. Desai, S. Gupta, B. Jose, S. Mukhopadhyay, K. Joshi, A.E. Islam, and M. Alam, "A comparative study of different physics-based NBTI models," *IEEE Trans. Electron Dev.*, **60**, pp. 901-916, 2013.

Some excellent lectures on NBTI, HCD, and TDDB can be found in Lectures 8-26 of the following online course:

[5] M.A. Alam, *ECE 695A Reliability Physics of Nanotransistors.*, *https: //nanohub. org/ resources/ 16560*, *2013*.

For good reviews of the current understanding of HCD, see the following papers.

[6] S. Mahapatra and R. Saikia, "On the Universality of Hot Carrier Degradation: Multiple Probes, Various Operating Regimes, and Different MOSFET Architectures," *IEEE Trans. Electron Dev.*, **65**, pp. 3088-3094, 2018.

[7] M.A. Alam, Y.-Pu Chen, W. Ahn, H. Jiang, and S.H. Shin "A Device-to-System Perspective Regarding Self-Heating Enhanced Hot Carrier Degradation in Modern Field-Effect Transistors: A Topical Review," *IEEE Trans. Electron Devices*, **66**, pp. 4556–4564, 2019.

[8] S. Mahapatra and U. Sharma "A Review of Hot Carrier Degradation in n-Channel MOSFETs – Part I: Physical Mechanism," *IEEE Trans. Electron Devices*, **67**, pp. 2660–2671, 2020.

[9] S. Mahapatra and U. Sharma "A Review of Hot Carrier Degradation in n-Channel MOSFETs – Part II: Technology Scaling," *IEEE Trans. Electron Devices*, **67**, pp. 2672–2681, 2020.

The original anode hole injection model for thick oxides and the updated AHI model for thin oxides are described in the following two papers.

[10] J.C. Lee, I-C Chen, S. Holland, Y. Fong, and C. Hu "Modeling and characterization of gate oxide reliability," *IEEE Trans. Electron Devices*, **35**, pp. 2268–2278, 1988.

[11] M.A. Alam, J. Bude, and A. Ghetti "Field acceleration for oxide breakdown-can an accurate anode hole injection model resolve the E vs. 1/E controversy?" *38th International Reliability Physics Symposium*, **35**, pp. 22–26, 1988.

Index

absorbing contact, 116
accelerated testing, 224, 231, 235, 241
access transistor, 179
accumulation region, 140
activation energy, 226, 230
active region, 210
activity factor, 20
AlGaN/GaN, 201
Anode Hole Injection (AHI) model, 237
apparent mobility, 118, 119, 123, 128
avalanche breakdown, 159, 160

back gate, 136
ballistic
    MOSFET, 100
ballistic limit, 114, 117
ballistic mobility, 100
bathtub curve, 221
beyond pinch-off region, 8
bipolar transistor, 51
bitline, 175
blocking voltage, 154, 157, 160
body effect, 136, 145
breakdown
    hard, 237
    soft, 237
breakdown voltage, 160
built-in potential, 145, 203
built-in voltage, 137

burn-in testing, 222
byte, 175

capacitance
    gate, 59, 87
    oxide, 59, 84
    semiconductor, 87
    switching, 18
channel, 2
    length, 4
    minimum length, 45
    resistance, 118
channel length modulation, 75
channel resistance, 66
charge
    mobile
        subthreshold, 86
charge carriers
    2D electrons, 79
    electrons, 9
    holes, 9
circuit convention, 4
CMOS, 10
CMOS inverter, 16
    activity factor, 20
    dynamic power, 19, 20
    noise margins, 17
    pull-down transistor, 16
    pull-up transistor, 16
    standby power, 16, 19
    static power, 19